Total Quality Management in Health Care

Dr Hugh C H Koch

TOTAL QUALITY MANAGEMENT IN HEALTH CARE

Longman Group UK Limited, Longman Industry and
Public Service Management Publishing Division
Westgate House, The High, Harlow, Essex CM20 1YR
Tel: Harlow (0279) 442601
Fax: Harlow (0279) 444501

© Longman Group UK Limited

All rights reserved. No part of this publication may be reproduced,
stored on a retrieval system, or transmitted in any form or by any
means, electronic, mechanical, photocopying, recording or otherwise,
without either the prior permission of the Publishers or a
licence permitting restricted copying issued by the Copyright Licensing
Agency Ltd, 90 Tottenham Court Road, London W1P 9HE.

First published 1991
Reprinted February 1992

A catalogue record for this book is available from the British Library

ISBN 0-582-07695-1

Typeset by Typestyles (London) Ltd, Harlow, Essex
Printed in Great Britain by Ipswich Book Co, Ipswich

To

Sue,

James

and

Emily

'The Health Service only exists because it has patients. It exists for its patients.'

'If perfection could be attained, it would not be worth having.'

Zen proverb

Contents

1 Introduction to Total Quality Management (TQM)   1
    Total Quality Management
    Quality assurance and TQM developments in the 1980s and 1990s
    Characteristics of TQM and control
        Quality leadership
        Companywide approach
        Continuous measurement and training

2 Preparing for Total Quality Management   13
    Clarity of NHS objectives
        Example 1 Trust mission statement
        Example 2 Pre-trust unit
    Phases of preparation for TQM implementation
        TQM audit: the first steps
    Diagnostic review
    Costs of poor quality
        What are quality costs?
        Why measure health care quality costs?
    Audit instruments for TQM
        Audit instrument I
        Audit instrument II
        Audit instrument III

3 Organisation and management of TQM projects   28
    Transforming the culture
    Creating a Total Quality culture in health care
    Making the change to TQM
        Quality policy statement: Hastings HA
    The change process
    Raising staff awareness
    Actioning quality improvements
       Key coordinating steering group
       Quality action teams
       Specialty departments
       Other initiatives

4 Information for Total Quality Management   34
    Quality and performance indicators

5 Standard setting — ensuring we get it *right first time*   41
    Participation in standard-setting
    Examples of standards set
    Monitoring of standards
    External standard setting

6 Clinical audit — improving care through audit   57
    Medical audit
        General methodology of audit
    Nursing audit
        Individual patient's nursing care
    Clinical therapy services
    Integrated clinical audit
    Conclusion

7 Communications: getting the message across to patients,
   staff, and teams   89
   Provider unit communication improvement
   Provider unit communication strategy
      Communication strategy
         Mission statement
   Communications audit: staff
   Communications skills training
      Communicating with patients
      Communicating with colleagues
      Who could use better communication skills?
      Learning better communication skills
   Communications improvement implementation plan: internal
      Team briefing
      Open forum meetings
      Unit newsletter and TQM information
      Teamwork enhancement
      Standards' setting, monitoring, and review
   Communication audit: patients
      Communications improvement implementation: patients

8 Training for quality — investing for the future   104
   Prevention of problems and errors
      Reporting and analysis of problems and errors
      Problem investigation
   Training needs
      Training method
   Conclusion

9 Evaluating the contract for quality   114

10 References   118

Preface

Sustaining and improving Quality of Health Care, within the NHS, has always been at the forefront of treating patients, both in hospitals and in the community services. Reorganisation of administrative boundaries and management responsibilities have occurred on several occasions, to facilitate a direct or indirect effect on Quality of Care and delivery of that care. The emergence of 'Total Quality' as an approach to organising, sustaining and improving Quality of Service offers considerable potential benefits to the Health Service in the 1990s and beyond.

TQM is an integral part of effective management, not an optional extra. It is a system which in general results in greater awareness of how performance in delivering health care can be explicitly measured against agreed standards to the satisfaction of both patients and 'purchasers'. TQM is a comprehensive approach, both cultural and technical, and must embrace all staff, all disciplines and all activities. This text, one of the first to address TQM and Health Care, offers a description of what TQM is, how the Quality culture can be developed, what the contributory technical components are and the benefits which can accrue from implementing this important organisational initiative.

Acknowledgements

I would like to thank the following people for their help, support and advice in the preparation and completion of this book:

Caroline Greenwood, Anthea Nicholas and Sue Alleyne, in deciphering handwriting and turning this into print.

Peter Wood, in showing continued interest in my ideas and approach, in theory and practice.

Alan Dearling, for his professional advice in completing this book.

My main Health Authority clients, who have offered me the opportunities to develop my ideas through implementing Total Quality Management.

My wife, Sue, and James and Emily, for monitoring how well this project has been progressing!

Hugh C. H. Koch

The publishers wish to acknowledge the individuals and organisations listed below for their kind permission to reproduce the following:

HMSO: John Oakland 1988 *Total Quality Management* Department of Trade and Industry (Figs 1.1; 1.2).
HMSO: Department of Trade and Industry (Fig. 1.4).
David Mathew 1990 *Management Team Effectiveness* GMTS Scheme London. (Management Team Effectiveness ALPTEC checklist).
Hastings Health Authority: *Investing in quality health care*, 1990 (Fig. 3.1).
Dr. A. Mason, Regional Medical Officer, South Western Regional Health Authority (Figs. 4.1; 4.2; 4.3; 4.4; 6.12).
Mrs. M. Lewis, Assistant Nurse Director, East Gloucestershire NHS Trust (Figs. 6.13; 6.14; 6.15; 6.16).
HMSO: *Measuring the Quality Package*, Department of Health (Fig. 6.1).
Royal College of Physicians Publications: A. Hopkins 1990 *Measuring the quality of Medical Care*, (Fig. 6.2) and *Medical Audit: A First Report*, 1989 (Fig. 6.7).
Dr. Nigel Offen, FRCS, British Association of Medical Managers (Fig. 6.6 and Template for quality (p.68)).
CASPE Research/Brighton Health Authority Department of Public Health *Summary of Generic Screening Criteria* RSCH/SEH Occurrence Screening Project (Fig. 6.22).
HMSO: NHSME *Nursing Care Audit* 1990 Department of Health (Fig. 7.1).

Despite our best endeavours, we have been unable to trace the authors for Figs. 6.17, 6.18 and 7.3.

Chapter 1 Introduction to Total Quality Management (TQM)

The issue of defining, measuring, monitoring, and improving the quality of health care and service has been addressed in various ways over time. Florence Nightingale when walking her 'patch' in military hospitals will have used her intuition and her experience to assess good and bad qualities of care. More recently, nurses, as highly trained professionals, would through their skills and competencies adapt and update their practice to meet patient needs. In the last decade, hospital hotel services, would as a result of competitive tendering requirements specify, or have specified for them, clear quality standards alongside quality and cost expectations. Throughout these activities, the issue of how standards are made 'explicit, operational, measurable and consumer-sensitive' has been crucial and variably addressed. The recent passing into legislation of *Working for Patients* and *Caring for People* is based on processes which will introduce market forces and the contracting process as a means to more accurately specify and then raise quality of health care and service. Over the next two to three years the recommendations in this legislation are likely to bring the NHS more in line with private sector companies, who have focused upon quality as an essential ingredient for successful development and achievement, developing quality processes in every aspect of their work. Today quality is a top issue in general business and specifically in the provision of health care in Britain.

General management of health care, when introduced via the Griffiths Recommendations (1983), placed the establishing of quality standards and review mechanisms on the agenda of all General Managers at Regional, District and Unit level. Individual Performance Review (IPR)/Appraisal was the vehicle, supposedly, to monitor the successful implementation of QA *(quality assurance)* across the acute and priority services alike. Considerable work and effort has been put into many services throughout the country to produce written standards of care and service and to become more patient-consumer sensitive. With the benefit of hindsight, two constructive criticisms of the Quality work in the 1980s are that:

1. Often QA was *delegated* by general managers to staff officers called QA officers (or equivalent). These highly competent people, usually from a nursing background, have in many cases sought to develop QA methodology in the absence of any structured training. They have attempted to manage the implementation of quality processes *on behalf of* their general manager colleagues. Though clearly achieved in part this is not true or effective quality management which has to rest on the general manager's *personal* agenda.
2. Quality assurance or management was not perceived explicitly as involving what would loosely be called *cultural* elements. Individual staff groups were, to some extent, involved in owning the standard setting process, but overall the drive to monitor and review on a regular basis the care and service offered in order that it could improve and be more acceptable to patients and carers was not seen as a Total Company Process ie one which, not only is *managed* by key senior individuals, but has implications for the action of *all* individuals in the service. Quality management was not *total*.

The recent 1990 legislation will facilitate the development of the contracting process in which *provider units*, whether they be NHS trusts, or Directly Managed Units (DMU), will be responsible for meeting specifications for services laid down in contracts with *purchasing agencies* (Commissioning

(Commissioning Agencies/DHAs, FSHAs). Increasingly, the ability of purchasers to obtain value for money, the providers' ability to *lead the local market* by providing the higher quality of care, and the general public's capacity to understand and expect effective, well organised health care will lead to the quality of care and service being an even more crucial component, if not the most crucial (alongside quantity, throughput and cost) in the NHS in the 1990s.

In most hospital and community services now, managers recognise that quality is a *survival* issue but one which is difficult to actively lead from the top. This is in contrast to successful businesses in industry. In a recent ODI survey, 76 per cent of top industry managers considered they actively help to improve quality in their organisations, compared to 38 per cent of their NHS General Management colleagues (Keyser, 1989). NHS managers are now very much aware of the need for quality improvement and acknowledge that the biggest problem is in making it happen, especially at times of considerable change and disruption.

The experience from industry worldwide shows that while individual efforts in quality assurance; standard setting within departments; customer care all produce good results, the highly successful companies are those dedicated throughout their organisation whether small or large, to total quality.

Total quality management

Errors, however defined, poorer practice and service, inconsistency and inefficiency in service operations are part of a lively and thriving service. They are also its scourge as they have a way of multiplying, creating problems in other parts of the company, leading to more errors, more problems and so on. As a result, people spend considerable time identifying these errors; correcting them or *patching them up*; apologising to individual customers. However laudable these activities, once the error has occurred, they have the quality very often of the 'closing the stable door' analogy.

About 30 – 40 per cent of effort in businesses in the UK is spent, and an arguable proportion wasted, on error detection and resolution. Alternatively, the benefits of ensuring work is going well and is always being done properly first time, are considerable in any business in terms of: greater consistency of high product quality; greater efficiency; lower costs; better market reputation and share; and better staff commitment and morale.

Total quality management (TQM) is a way to manage the many processes which ensure these quality issues pervade and infiltrate every aspect of an organisation to improve its effectiveness and competitiveness, and ability to flexibly adapt to new conditions. Worldwide it has been applied in both the manufacturing and service industries and is slowly being brought into the major UK public sector services. It involves whole businesses becoming quality-sensitive and organised, in every department, every activity, every level, and involves every individual. Quality management, like certain advertised lagers, must reach every single part (of the organisation)!

In business terms, over time, *revolutions* have come and made their contribution, whether they have been for example, *Industrial* (19th Century) or Computer (1970-80s). We are now in the centre of the *Quality* Revolution where there is a noticeable momentum in many businesses to identify and improve the quality of their product service. Continuous quality improvement and cost containment are essential for successful business activity.

Total quality management helps an organisation to achieve active processes as in Table 1.1.

Table 1.1. TQM processes

- Focus on the needs and expectations of its market and its consumers
- Achieve top quality performance in all areas of its activity (product service and internal processes)
- Install and operate procedures, simple and complex, necessary for the achievement of top quality performance
- Critically and continuously examine processes to reduce and remove non-productive activities, inefficiencies and waste
- Develop and monitor measures of performance, set standards against which this performance is measured, and set required improvements
- Understand and develop an effective communication strategy
- Develop a non-hierarchical team approach to problem solving with delegated responsibility for change
- Develop good procedures for communication and feedback to staff at any level of good work
- Review continually the above processes to develop the culture for never-ending improvement

Source: Oakland, 1989

Private and public sector business increasingly has to address quality issues in relation to competition and holding market share. In health care, the issue of cross-boundary flow of patients being referred to hospitals and services *outside* the neat NHS boundaries of Trusts, Units, Districts and Regions is a key one to address within this context. Consumers place a higher value on the quality they receive than on loyalty to their local provider of homebased products. Price and geographical accessibility are relevant but no longer the major determining factors in consumer choice. Consumers have greater choice and have increasing expectations. TQM helps organisations to address the many aspects of quality and keep in touch with their consumers' views and requirements — all this is relevant, as will be shown in later sections, to health care provision in the UK.

The concepts of TQM are very simple and based on commonsense. Any organisation requires processes for ensuring the service it provides is acceptable and wanted by its market. These several processes are complementary and *should* build up to an uncompromising commitment to quality and quality improvement.

In a recent Department of Trade and Industry publication (1989), John Oakland characterised TQM figuratively as in Figure 1.1.

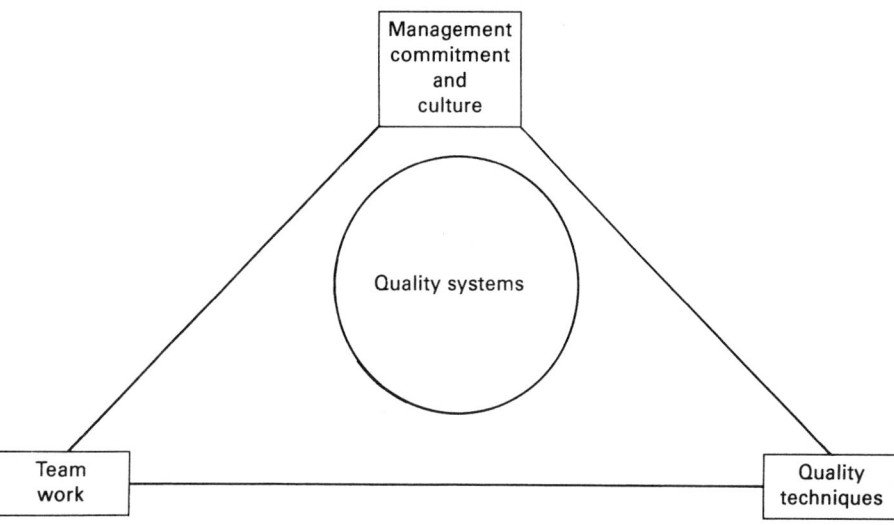

Figure 1.1. The TQM Model

Source: John Oakland, 1989 (Department of Trade and Industry)

This ensures the implementation of management commitment to quality and the development of a quality conscious culture; provides quality systems backed up by specific techniques for addressing quality maintenance and improvements; and crucially harnesses and develops the teamwork at several levels to ensure commitment to creative problem solving.

These aspects are often applied separately or as *flavours of the month* in companies and organisations. However, TQM involves a systematic and structured approach to the launch of quality improvement as depicted in Figure 1.2 (Oakland, 1989).

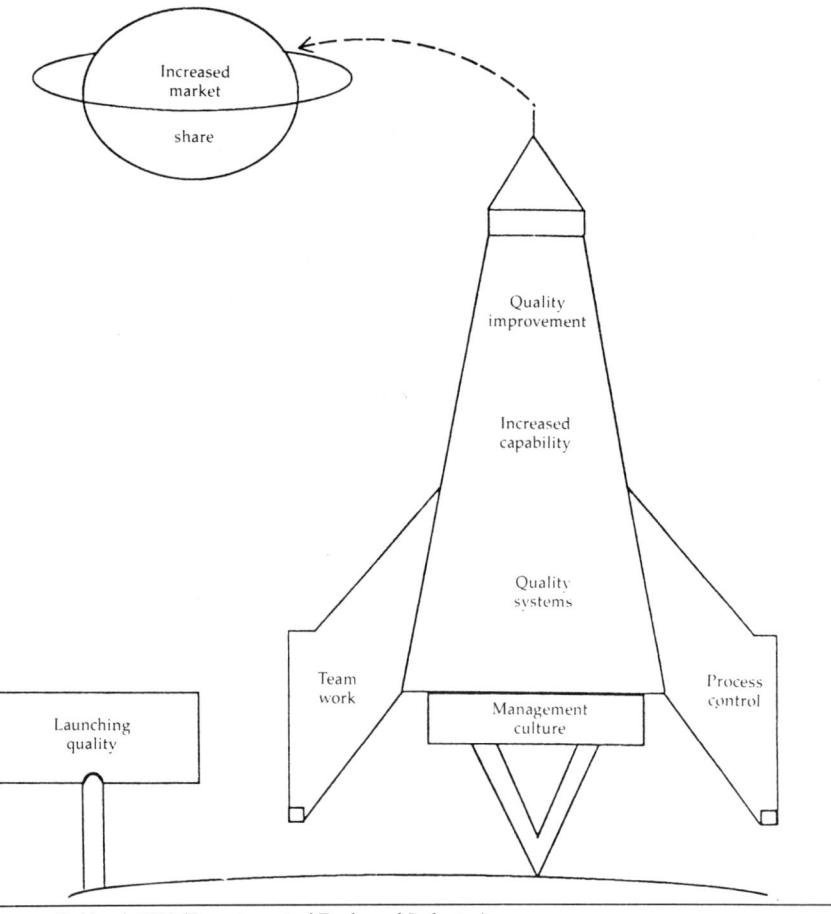

Figure 1.2. Launching quality improvement

Source: Oakland, 1989 (Department of Trade and Industry)

TQM enables, or could enable, a Hospital, Unit or Trust to improve its ability to meet and exceed its patients' requirements through an organised approach to monitoring, reviewing, and enhancing the quality of care and service offered adopted by *all* staff of *all* grades.

Key aspects of this approach are:

1. Commitment by all to quality improvement
    Developing a culture of continuous searching for improvement and positive, healthy change.
    Truly *companywide* quality culture in which *doing it right, first time* is almost second nature.
2. Meeting and exceeding customer requirements
    Identifying ways to establish what patient/consumers want and improving health services abilities to respond to these.
    Exceeding requirements of availability, delivery, reliability, maintainability and cost effectiveness.
3. Understand internal customer requirements
    Satisfying internal staff suppliers and customers.
    Identifying the internal quality chain and preventing operational *breaks* in these chains.

4. Maintaining quality of *Design* of service and conformance to *Design*
   Establishing, maintaining and reviewing key quality standards of *all* services.
   Ensuring standards are explicit and measurable.
   Ensure agreed standards are adhered to and that non-conformance is identified and addressed.
   Ensure customer views are central to standards set.
   Understanding the costs of non-conformance and not accepting them.
5. Ensuring ownership throughout the service for quality functions
   Encouraging staff at all levels to identify and facilitate quality improvement.
   Promote the partnership in quality between internal and external customers and suppliers.
   Encourage ownership of planning, managing, auditing, and reviewing quality standards and systems.
   Encourage staff's inherent motivation and good sense to improve the service.

TQM means . . . . .

- Putting the patient and carers first.
- Being fully aware of patients' expectations and needs.
- Satisfying each patient and satisfying them first time.
- Recognising the costs of poor quality.
- Supporting and encouraging every member of staff in every hospital.
- Encouraging staff loyalty to their department or ward and to their hospital, unit, district or trust.
- Encouraging enthusiasm, knowledge, and skills in staff to help them deliver higher quality and cost effective health care services.
- Encouraging professionalism and expertise among all staff.

Figure 1.3. Key aspects of TQM in health care

These can be combined with the overall strategic model (Figure 1.1) and will form the basis for ensuing sections of this book as shown in Figure 1.4.

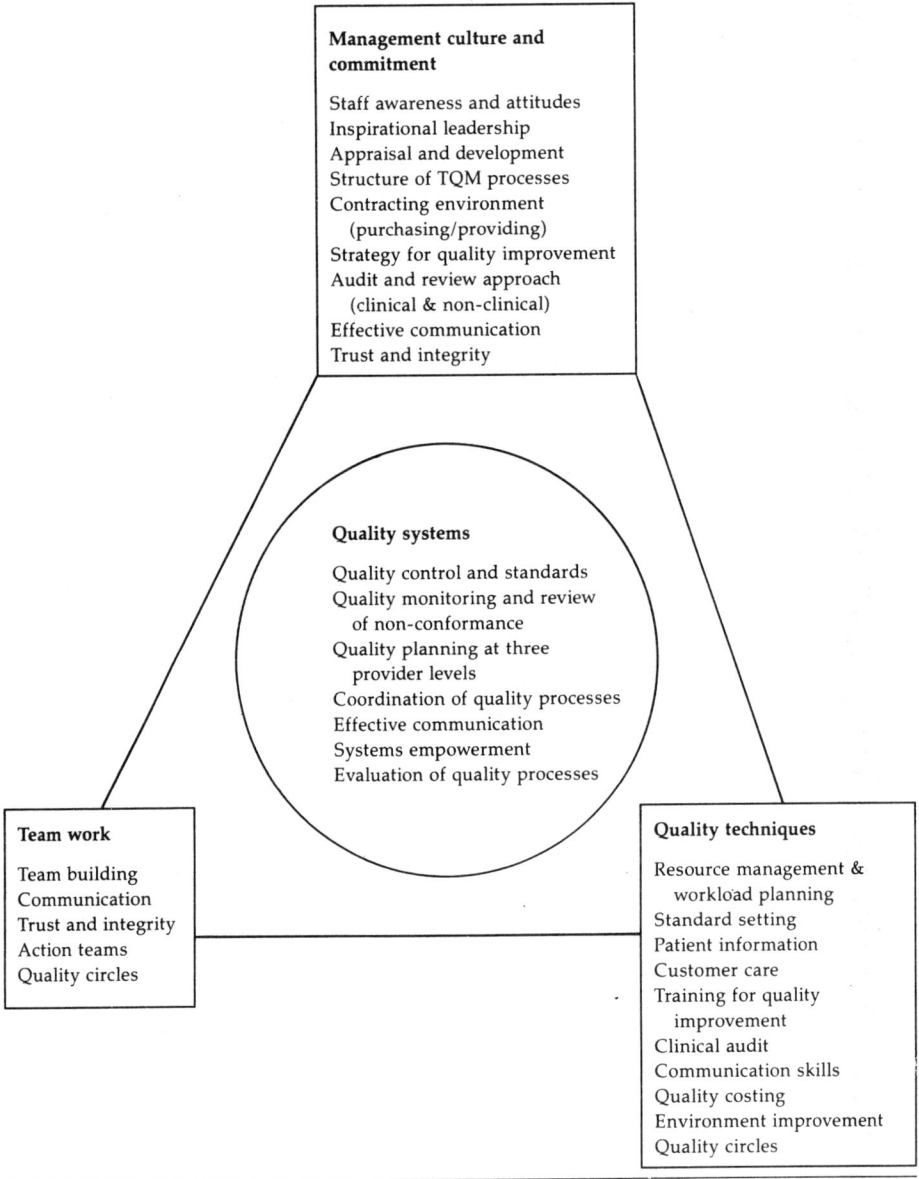

Fig. 1.4. TQM strategic model

Source: Department of Trade and Industry (adapted by Hugh Koch)

Confusion can sometimes arise in many quarters of health services concerning understanding, allegiance to, and integration of the following major initiatives:

1. General management
2. Business planning
3. Resource management
4. Total quality management

At the risk of being oversimplistic, many managers both inside and outside the service see these four initiatives as being complementary if not coincident in that they all represent *good management*. Inevitably, each has its own emphasis and unique elements in terms of, for example, *personal* accountability[1]; Market research and financial forecasting[2]; information technology and clinician involvement[3]; and customer-responsiveness and audit[4]. However, they all involve corporate ownership of providing high quality services within the boundaries of what is affordable and possible in terms of workload and throughput. Health care in the 1990s must be worth purchasing. As described more fully elsewhere (Koch, 1990), the issue of quality is paramount because:

> The success of hospitals is likely if quality is taken seriously. With increasing customer choice and purchasing power, cost efficient and high quality services in hospital and community settings will attract variable cross-boundary flow away from services of poorer quality. In the near future, differentials in quality of care will result in differential financial viability and capital development.
> Rising expectations of customers
> A crucial spur to activity in the NHS is the increasing awareness of the public as to how quality of care can vary and why. Acceptability of service is becoming a major issue. The public will rightly expect the health care purchased on their behalf to be of identifiable, acceptable, and rising standards.
> Concentration by purchasers on quality conscious providers via the developing relationship between purchasers and providers, the contracting process will encourage providers to identify quality — as well as cost — and quantify improvements in order to maintain or increase its *market* share and hence predict sustained growth of services.
> Cost advantages of providing quality
> All health care professionals are aware that poor quality in hospitals and community services costs money. *Putting things right* takes time and effort and additional resources of both capital and revenue. These *costs* are usually hidden and often forgotten or ignored. To identify these hidden costs and redress the problem can lead to releasing resources for reinvestment.

Quality management in health care is not new! Ever since the NHS was set up, and even before this, people working in hospitals and in the community have been striving in different ways to provide good care and service; to do their best; and to look for improvement. TQM ensures these efforts are harnessed, coordinated and apply to *all* aspects of the complex and diverse services which make up the NHS. *Quality* will continue to increase as it is both *managed* and *total*.

Quality Assurance and TQM Developments in the 1980s and 1990s

Throughout all organisations worldwide, the 1980s saw an unprecedented growth of emphasis upon quality and quality processes. To many providing services and purchasing services, *quality* is equal to, if not more important than price in how decisions are made. This change in emphasis is a rapidly growing trend in business of all kinds. The reinforcement that the Japanese car manufacturers received in the 1980s for their attention to detail and good quality via increased market share in the western marketplace is proof of this trend and its successful outcome. Similar trends occurred in electronics, telecommunications, chemical industry, and service industries including hotels, airlines, banking; with preliminary steps in health care, social services, and police forces.

Much of the initial momentum for quality improvements relied on market force and necessity. This was gradually complemented by awareness of how quality improvement could result in lower cost as well as more efficient utilisation of human and non-human resources such as equipment, land, buildings, supplies.

Marketplace and cost trends, as Feigenbaum (1988) noted, will result in a period of very active quality competition in almost all types of industry and service in the next decade. There is, according to most informed business leaders, a great variation in the readiness or organisations to meet the challenges they will inevitably face in terms of either market or cost in the

1990s. Quality awareness is important and *relatively* easy to achieve — the execution of quality processes with informed implementation of quality improvement throughout any one organisation is a slow and sometimes arduous process, which most organisations have only just begun. There are clearly lessons that health care providers and purchasers can learn from how Total quality control has been introduced in non-health care fields during the 1980s and into the '90s.

Characteristics of TQM and control

Many of the most successful and well organised companies are recognising the importance of TQM and are preparing themselves for the inevitable demands for higher quality that they expect in the future. Business planning is being based more and more on the assumption that to exceed customer requirements at level, or even reduced, cost can best be achieved by the emphasis on quality management and improvement.

Characteristics of TQM in the private sector worldwide have been:

1. *Company-wide quality improvement* achieved by involving the help and participation from all staff in an organisation — not just senior managers or quality control/assurance officers.
2. *Quality improvement applies to all services.* It is not restricted to the literal production worker, it applies to all aspects, all functions of all industries — manufacturing, service, private and public sector.
3. *Quality improvement must be systematic,* extending company-wide and not restricted purely to technical functions.
4. *Technology does aid quality improvement,* in that modern quality design techniques, computer aided measurement and control and modern communication processes all help the technical aspects of quality management.
5. *Quality must satisfy the customer,* in that quality improvement is essential to meet and exceed customer, both internal and external, expectations and should not be determined purely from a non-customer-sensitive professional view.
6. *Quality systems must support innovation.* Quality is part of everyone's job in a company. Quality processes must be organised in ways which support excellent performance of individuals and teams.
7. *Customer-oriented quality strategy.* A clear, unambiguous widely accepted strategy plan and system for quality maintenance and review.

Increasing sales growth, market share and lower costs are being achieved via the introduction of TQM philosophy and plans in Europe, USA, and the Far East. Table 1.2 identifies a selection of companies/industries which have introduced and benefited from the application of TQM approaches.

Table 1.2. Types of companies with TQM approaches

	USA	UK/Europe	Far East
Manufacturing	Electrical Semi-conductors Glass works Motor companies Tyre companies Telecommunications Computer	Telecommunications Rank Xerox Electronics Chemicals Construction industry	Car manufacturers Engineering Packaging
Service	Health care Banking Hotels Airlines Transport	Hotels Banking Airlines Transport Department stores	Hotels Airlines Department stores
Public Sector/Service	Health care	Royal Mail Health care Police force Social services Local authorities	

TQM and its related management processes have been widely written about throughout the world by authors such as Peters and Waterman (1982); Crosby (1979,84) in the USA; Ivan (1979,80) and Demming (1982,86) in the USA and Japan; and Oakland (1989) in the UK.

Quality management and leadership developed in companies working in areas shown in Table 1.2 is based on programming unrelenting attention to quality and total quality control into the running of their businesses through three main processes or components: quality leadership; company-wide approach; and continuous measurement and training.

Quality leadership

Clear plans to produce quality goods and the resourcing required to meet these goals in terms of manpower and capital expenditure mean an explicit, strategic and policy commitment to low or zero defect rates, that is *getting it right first time*. Leaders must ensure that continuous improvement is the norm and that consumer expectation of services and products will, and should, always rise. Quality is ever rising upwards: the more successful a *product* becomes the higher the quality must be if the organisation is to grow.

Company-wide approach

All the necessary quality processes and actions must be implemented throughout entire companies, not just in isolated departments or atypical centres of excellence. In all successful *quality* companies, quality improvement is seen as a *sine qua nos* of management, production and service, and must be applied by each part or function of an organisation. Managers and supervisory staff take the lead in systematically attaining and maintaining improvements. It is not failure-driven but prevention-driven. It is not static, but constantly developing new approaches to products and assuring their quality. Product development and quality systems development go hand-in-hand.

Continuous measurement and training

Training and continued updating/education has been successful where it is specifically job-related. General industrywide training is not successful. The engineer in design fields and manufacturing will clearly have different training needs to enhance the quality of his/her work. Having said that, this has been shown to be even more effective when an integral part of total quality management, with widespread employee understanding of widely agreed quality processes and disciplines. The main motivation for this rests with the high results in industry and business from genuine advances and commitment to quality leadership. Return on investment in strong total quality programmes in large firms worldwide has been high. This has resulted from three things:- improvements in quality costs ie the full cost of failure of controls; improvements in productivity, via reorienting *repair* departments to *productive* departments; and higher sales/market share. Market share correlates with market quality. Successful sales occur for products and services with high quality/effectiveness and low quality costs.

Total quality management is becoming a central requirement throughout the world for manufacturing and service industries, both in the public and private sectors. It is a way of managing, by achieving customer satisfaction and lower costs together. Outside the health care industry, companies are finding the TQM path, however it is labelled, a very difficult and frustrating one to begin with. It is being implemented at variable rates and with varying commitment. This is accompanied by lower tolerance for the time and cost of produce service failures. Therefore for both economic and social reasons quality improvement is central to the healthy development of all businesses.

Health care professionals can be reassured that those working in other industries and businesses have found and still are finding that sustained quality improvement is both essential and difficult. What has been occurring in the NHS over the past decade in terms of specific quality initiatives which now require sustained commitment to integrate these into an overall 'provider' approach is not dissimilar to what many other companies around the world have also found.

Traditionally, quality of health care and its maintenance and improvement was the main responsibility of the professions ie doctors, nurses, etc. On the basis of their training and qualifications they controlled the *implicit* standards for care, backed by a variety of relevant clinical *checks and balances* such as case discussions, morbidity and mortality meetings, and postgraduate training events.

This professional quality control was supplemented by the roles and responsibilities of health authorities in their various forms, which resulted in part from difficulties and crises in health care in the 1960s (eg Hospital Inquiry Reports, DHSS, 1969). In addition, external advisory groups (Hospital Advisory Service; National Development Group for the Mentally Handicapped; Mental Health Act Commissioners) complemented the internal quality control of authority members and the professionals themselves.

The precursor to Griffiths-style general management was introduced following the 1974 reorganisation with clearer roles for service administrators and clearer responsibilities for efficiency and quality service (DHSS, 1972). At approximately the same time in the early 1970s, greater emphasis was placed on the role and voice of the patient — the consumer — via a clearer process for investigating patient complaints both within the service and external to the service with revised hospital complaints procedures and the establishing of the NHS commissioner service. In addition, the establishment of Community Health Councils (CHCs) gave, potentially, a stronger and more coherent voice for the users of the service. The ability of the CHCs to adequately represent and communicate consumers' views has been the subject of considerable debate both by CHCs themselves and by the services they aim to help improve.

Quality improvement in hotel services was inadequately addressed with variable success as a result of attempts to increase efficiency and cost-effectiveness during the early 1980s. The putting out to tender of many support services (eg domestic, catering, portering) began to introduce the competitive element into service efficiency and quality. The requirement for hospital *in-house* services to submit tenders, alongside private firms, every three or five years, meant that the contract specifications — so much in vogue now as part of 'Contracting – 1990 style' had to include explicit statements as regards costs, quantity *and* quality. On the negative side, some districts saw a worsening of labour conditions and pay rates within these services. However, on the positive side, considerable savings were identified during the 1980s which purportedly and variably went to improve direct clinical services, and secondly the quality of hotel services was made far more explicit and accountable, and in some cases led to improvement of standards.

The circular *Implementation of the NHS Management Enquiry Report* (June, 1984) was issued as the government's response to the NHS *Management Enquiry Report* (June, 1984) which described a defused general management function incorporating responsibilities for performance control and quality. It was intended that general managers would move towards greater overall control though devolution; coordination of resources and their management; corporate direction; plus the more effective management of internal and external relations. Some of the early successes and aspirations of general managers are cited and illustrated in Koch (1988).

As a result of the ministerial review in 1988 into the NHS, two white papers emerged. They were *Working for Patients* outlining ways to improve patient choice, service efficiency, and quality via the introduction of a contracting process (plus greater emphasis on organised clinical (medical audit)); and the second *Caring for People* addressing the funding and coordinating arrangements for community care services. Both papers passed into legislation in 1990, predicting improvements in service quality in acute medical care and priority services.

There are in fact many signs of excellent progress in the NHS on quality and approach to maintaining and improving quality. The introduction of general management in 1984 led to a greater coordination of specific quality assurance activities eg standard setting; customer satisfaction; and patient information upgrading. Simultaneously and independently of general management initiatives, the clinical professions (medical, nursing, professions allied to medicine (PAM) continued variously to audit their work with some positive benefit to patient care. In other words, some technical components of total quality were already receiving attention.

With the redesignation of responsibilities of the NHS management executive, and its relationship with its political master (and at that time mistress!), plus the very significant appointment of a highly qualified and credible service manager to its direct executive position, further progress has been stimulated on quality issues from the centre.

DHAs have been instructed *(EL (89) MB117)* to ensure that their units develop systematic and continuing quality review experience, using a format and content determined locally, but consistent with national and regional policies. They should include provision to monitor all aspects of quality of patient care and service, including outcomes.

Specific attention will be paid to:

- Medical and clinical audit
- Reducing waiting times (outpatient and inpatient)
- Specification of quality elements to contracts
- Measurable criteria or standards of care/service
- Improved appointment systems
- Information to patients
- Reception and public area arrangements
- Customer feedback on strategies
- Improved environments (eg Accident & Emergency)

To assist the progress on these several issues, support guidance and funding have been forthcoming from the Department of Health (DoH) during 1989 and 1990 across a range of significant quality areas. In addition, the DoH funded centrally *demonstration sites* to pilot the following:

- Outpatient Departments:
    Queen Elizabeth Hospital (Gateshead); Nottingham City Hospital; Fazakerley Hospital (Liverpool); Leeds General Hospital; St Georges Hospital (Tooting); Torbay General Hospital.
- Total Quality Management
    Canterbury and Thanet; Cheltenham; Doncaster; E Berkshire; Grimsby; Liverpool; Merton and Sutton; Milton Keynes; Mid Staffs; SE Staffs; Northallerton; SW Herts; Trafford; Waltham Forest; W Dorset; Winchester; Worthing.

The TQM demonstration sites were funded in 1989. They varied from specific QA type initiatives through to full blown TQM implementation programmes the latter received further funding in 1990 alongside a small number of additional TQM projects.

The Total Quality Management (TQM) approach was defined as putting the needs of the patient/customer at the centre of every activity, including staff in departments and at all levels of the organisation, with the crucial commitment and involvement of senior management.

The TQM approach has been summarised already in Figure 1.4. Subsequent sections will address the preparation for TQM; the organisation and implementation of TQM; and training for an evaluation of the TQM approach to health care delivery in the NHS.

Chapter 2 Preparing for Total Quality Management

Implementing a total quality programme requires active, very visible and effective leadership from all levels of management and supervising staff, starting with the provider Unit Chief Executive/Unit General Manager (or District General Manager in District Trusts or Districts with directly managed units (DMUs). If staff at senior or middle management levels are uninvolved, less than totally committed, the TQM programme will be limited in its effect of the quality outputs of care and service to the patient. The first step in establishing the ground for TQM is clarity of vision concerning the business objectives of health care locally, a thorough assessment concerning quality in the service; an understanding and identification of the customers, both internal and external, and existing quality management systems. Alongside these areas, it is necessary to assess the prevailing managerial culture and style with its likely effect or outcome of service quality.

Clarity of NHS objectives

Through the leadership policy of the Management Executive as outlined in EL(90)MB/77 and the implementation and quality management of its Chief Executive/Unit General Managers, and the purchasing strategies of the commissioning agencies, the NHS has the overall goals of:

Securing, within available resources, significant improvements in the health of the population through delivery of services providing health promotion, prevention and diagnosis, high quality care and service.
Ensuring effective, efficient and economic care which is relevant and acceptable to patients, and is *accessible*.
Ensuring support for staff providing these services.

and achieving them by:

Developing and implementing strategies to address the above goals.
Developing and supporting effective provider unit management and encouraging delegation.
Developing range and quality of training.

In the five years, 1991-96, the medium term aims are to ensure that the quantity, quality and effectiveness of health care services are improved through the placing of contracts, and the changes within the NHS and Community Care legislation following the implementation of three white papers *Working for Patients, Caring for People,* and *Promoting Better Health.* Alongside these objectives, effective control of financial, human, and capital resources/assets, plus securing and retention of appropriately skilled staff will be required. All this will require the maintenance of the recent momentum within the NHS towards increasing the effectiveness, efficiency, economy, and consumer responsiveness with which the NHS as a whole, and each individual provider unit, uses its revenue, capital assets, and human resources.
Translating this into operational *reality* requires:

Encouraging providers of services to respond more sensitively and quickly to individual patient needs.
Provide choice.
Focus attention on outcomes of clinical interventions.

Devolve more responsibility to doctors, nurses and local managers.
Contracts which specify quality and service standards.

Quality will be a decisive element in the contracting process between *purchasers* and *providers*. Providers will need to ensure their services are of sufficiently high standard to attract contracts from commissioning agencies, and will need to maintain these standards in order to keep contracts.

In very practical terms, operational issues of resource management (and *Better Patient Care SWRHA: Resource Management Strategy);* medical audit, clinical audit, contracting, and quality management are the four key areas of change that can lead to tangible changes and improvement in services and service quality.

The changes emanating from these initiatives provide a very challenging, if not awesome, management agenda. Within the next five years the health service will undergo a radical change. The large acute hospitals which anticipate and manage these changes successfully and effectively will place themselves in advantageous *market* positions *vis-a-vis* their competitors (where they exist).

Whether the new *providers* are labelled as NHS trusts or directly-managed units they will base their plans to achieve the 1990-95 NHS objectives, on an overall business planning philosophy — an approach which contains the following elements:

1. Underlying philosophy, mission statement and core values.
2. Plans for improving quality.
3. Leadership and management philosophy and structure.
4. Financial strategy including service development and contracting, pricing policy, income predictions, revenue and capital expenditure plans.
5. Personnel strategy including manpower and staff development plans; pay determination and conditions of service plans.
6. Information systems strategy including support staff arrangements and IT.
7. Estate management strategy including estate rationalisation, capital values and plans, property condition and required information systems.
8. Service delivery plans including summaries of current resources and workloads and changes predicted; policies for joint work by managers and clinicians in determining pattern of services.
9. Market research and consumer responsiveness.

All nine of these elements involve quality of some aspect of health care delivery. TQM is not separate from underlying plans, financial strategy, or information systems strategy. It encompasses them as the overall management philosophy style and approach to planning, delivery and evaluating health care services. Returning to the first element, however, a provider unit, like its predecessor units, should have an overall philosophy based on certain core values which are explicit and agreed by the majority, if not all staff. Two examples are given below of mission statements, the first from a first-wave trust application; the second from a pre-trust unit:

Example 1: Trust mission statement (East Gloucestershire Trust application, 1990)

Objectives   To provide a choice of comprehensive, high quality health care at hospitals in . . . ., at the associated community hospitals, through a range of community facilities and to patients in their own homes. Services will be free, accessible and will meet specified standards of quality as flexibly as possible.

Patients   Patients will provide the trust's livelihood and will be put at the forefront of thinking in all areas of activity. The aim

	will be to develop and maintain excellent working relationships with all general practitioners with whom the trust will work for the benefit of patients.
Scope	A full range of services will be provided to embrace health promotion, prevention and screening programmes as well as diagnostic and treatment facilities, in-patient and domiciliary services.
Quality	Only top quality will be acceptable. Information to patients will be improved. A Service Review Group will be established which will enable purchasers, general practitioners, patients, their relatives and the general public to communicate their views of standards of care provided.
Staff	The trust's most precious resource will be its staff. The trust will provide an open and challenging environment in which staff of all disciplines will be able to develop their abilities to the full and contribute at all levels to maintain excellence of care. The trust will be a fair employer.
Finance	Financial resources will be earned on merit. The trust will strive for efficiency and cost effectiveness. High quality services will be provided in return for a fair price. Prudent management will maintain financial stability and create surpluses which will allow a steady improvement in quality and quantity of services.
Development	Opportunities will be sought to be innovative, consistent with the ever-widening scope of medical knowledge and proven benefits. There will be a flexible response to meeting new needs.
Ethics	The highest ethical and professional standards will be applied across the trust's activities. The trust will be entirely straightforward and deal openly with patients, purchasers, suppliers, government departments and the community which it serves.

Example 2: Pre-trust unit (Cheltenham Acute Unit, 1989)

Mission statement

1. To maintain and improve the quantity and quality of acute health services available to all residents in . . . . within the available resources.
2. To increase and develop resources and level of care by releasing and generating income.
3. To facilitate development of professional expertise and services continuing to foster a positive and action-orientated approach to placing the *patient first*.
4. To inform and involve staff in appropriate activities.

Core values underlying mission statement

Six values are hopefully shared by all staff in the Unit as contributing to achieving the core purpose above. They relate to effectiveness, efficiency, quality enhancement, accessibility, consumerism, and positive teamwork, and are:

1. The main goal of the Unit is individual patient care of the highest effectiveness within the resources available and state of current practice and research.
2. Available resources, both human and financial, should be used as efficiently as possible ensuring the highest cost-efficiency whilst maintaining agreed levels and quality of service.
3. Standards of all services should be set and regularly monitored with annual targets for quality improvements.

4. Services should be as accessible both geographically and temporally as possible to patients requiring them.
5. Patients — our *consumers* — should wherever possible be *put first* in terms of the services provided, the environment, the information provided, and the way in which they are communicated with.
6. Good working relationships between staff and **all** disciplines fostering trust, mutual commitments, interpersonal integrity and recognition are essential. Views of staff at all levels should be elicited and included in the planning and implementation of the Unit's activities.

Once the overall *Umbrella* for the provider unit's approach to care has been established, one of the next steps in business planning is to establish how the prime objective of any unit — the provision of the highest standard of health care affordable — can be achieved. Such a policy would include:

Philosophy of management towards quality:
    Quality assurance, TQM
    Elements of quality eg Maxwell's (1984) dimensions of:
        1. Access
        2. Relevance
        3. Effectiveness
        4. Social acceptability
        5. Efficiency
        6. Equity

Measurement of quality:
    Market research
    Application of external standards
    Independent observations
    Audits
    Use of performance indicators

Current quality achievements:
    Quality for patients
    Quality in management
    Medical and clinical audit
    Health records quality
    Training in quality
    Complaints analysis

Future aspirations and plans:
    TQM framework
    Encouragement of TQM culture and commitment
    Incorporation, under TQM, of the following:
        1. Resource management
        2. Clinical audit
        3. Standard setting and review
        4. Patient information and communication
        5. Staff information and communication
        6. Consumer (patient and staff) feedback strategy
        7. Training for quality improvement

Specific reviews of:
    Control of infection
    Radiological protection
    Drugs and therapeutics
    Nursing policies and procedures
    Mental Health Act provision
    Admission and discharge policies.

Phases of preparation for TQM implementation

In preparing for implementation of TQM, two major phases are necessary. First diagnosis and gaining senior management commitment; and secondly establishing support structure and communications as in Figure 2.1. Phase I diagnoses the provider unit's present performance in terms of quality by:

Finding out what the *customers* think about the quality of care and service currently provided.

Increase awareness by all staff of need for quality improvement and identify key areas to address.

Establish and reinforce commitment at several management levels to improve quality throughout the unit.

Foster an appreciation of advantages of departments working together.

Establish need for standard setting, monitoring of performance and service improvement.

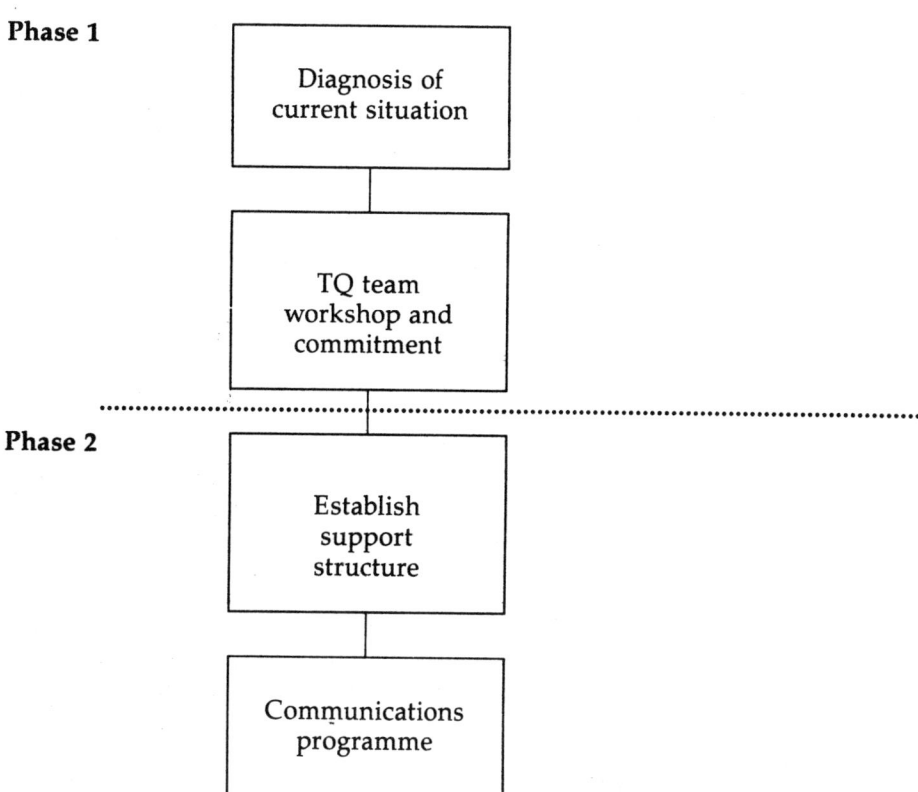

Figure 2.1. Preparing for implementation of TQM

This audit is followed by meeting/workshop with unit management board and senior clinical representatives to share and agree the *diagnosis* and possible *treatment required*.

Phase II takes the proposed programme forward to all the staff via a major communications programme to ensure they are all aware and committed to the total quality programme. By necessity this includes interdisciplinary workshops for heads of departments and services, as well as briefing sessions for all other staff by heads of departments. The aims of this communications programme is to *inform* and establish **views** on the programme.

Following these two phases, the TQM programme implementation can begin.

TQM audit: the first steps

The purpose of this pre-TQM and subsequent audit is to determine the degree of implementation of any quality systems and the degree of compliance with explicit standards. Such an audit helps to:

Measure quality systems.
Detect errors and trends.
Ascertain whether care and services are under control.
Identify areas where particular attention is needed.
Judge whether quality management plans are adequate.

TQM is a corporate approach to managing the business of health care delivery, and as such includes within it all the elements of quality, quantity, and cost and many subcomponents such as: standard setting, audit and training. It is potentiated by the presence of an organisational culture promoting the values and attitudes of consumer-sensitive individualised patient care of the highest affordable quality and effectiveness.

It is clear that TQM in health care is a long journey taking several years rather than an intensive *project* with a clear beginning and end. Phrases indicative of this approach are *quality ball on the hill, continuous improvement* and *road to excellence*. Many provider units are preparing for TQM as an extension of quality assurance initiatives in the 1970s and early '80s, and as a response to contracting for services.

Health services considering TQM need to identify their current position in terms of quality and delivery before taking further steps in improving quality systems. They need to carry out a management audit of their present situation to establish a baseline against which to measure progress in succeeding months and years.

This raises very interesting issues in that to audit, there must be a reasonably clear concept of what is the desired service or system. Because TQM is so new to health care professionals and managers, and despite similarities to manufacturing and service industries, there are no off the shelf audit instruments available. It means that in devising health service specific auditing tools, an uncompounded and organised view of TQM is essential.

It is at this stage that the quality issue gets even more interesting. Quality assurance officers, nurtured on QA methodology and general managers, reared on Griffiths I, resource management, and recent legislation on contracting converge to define: what is or is not TQM?. Pertinent and recurrent questions are:

Are quality assurance and TQM different?
Are TQM and resource management complementary?
How does quality relate to quantity and cost?

Diagnostic review

To undertake a diagnostic review, it is essential to do the following:

Identify current quality initiatives within the unit and the extent to which these *good practices* are generalised to *all* other similar service departments in the unit.
Gather information on how the unit is seen in terms of the quality of care and service by staff and patients/relatives/carers/GPs.
Design an action plan(s) based on the information gained in this diagnostic phase, aimed at increasing consistency throughout the unit in terms of good practice, reducing wasteful activities and producing a sustained and measurable improvement in quality of care and service in the unit.
Develop the competence and capability of managers and staff within the unit to maintain and progress action plans on quality in their own areas of influence.

The methodology involved in any diagnostic review of service quality includes five main elements:

1. Formal and informal structured interviews
    Carried out by the unit quality assurance officer, internally (but preferably **not** senior management) or an external management consultant.
    Held with all groups of staff throughout the unit to obtain a cross section of views (senior managers, middle managers, ward and unit managers, clinicians and GPs) concerning:

1. Current quality of service and assessment of performance (see later section)
2. Existing quality shortfalls
3. Potential opportunities to increase quality
2. Assessment of commitment to TQM values and management style
Via informal interview or checklist/questionnaire circulated to staff to determine the extent to which they believe in and are committed to TQM values, and the management style emanating from these values, and also the extent to which they feel the unit promulgated these values.
Values assessed are shown in Table 2.1.
3. Audit of quality systems
Selected systems and procedures purporting to support quality maintenance and improvement are analysed to assess strengths, weaknesses and potential for improvement.
Existence of service standards and monitoring processes are reviewed.
Quality improvement initiatives in each profession are identified and reviewed critically.
4. Problem tracking
Major problem areas, some new and some *well known* are identified with in depth analysis of causes and maintaining factors in the continuance of these problems.
Use is made of the *patient trail* model in which the *path* through a service which is taken be a patient is followed on paper or by discussion and pertinent questions raised concerning quality of care and service at different stages.
5. Quality costs
Amount of time spent in *wasteful – inefficient* activity for staff or patients is assessed and the relative costs in non-cash releasing or cash releasing terms is identified.

Costs of poor quality

At the preparatory stage of a health care TQM programme, an essential feature is to attempt to measure and *track* the cost of quality. For those managing and controlling the costs of services with special concern for the costs of both poor and good quality, the following needs to be available:

All quality costs, presented in a way that is relevant and face-valid;
Costing information for managers about what factors drive costs up or down significantly;
Assessment of how crucial poor quality costs really are in terms of the overall budget.

This financial aspect of quality is essential for a provider unit to perform effectively in the developing competitive market place.

Quality related costs in health care delivery matter because they are invariably large, but are often unmeasurable and uncontrolled. Nearly all *quality costs* are concerned with appraisal and failure rather than prevention. Failure costs are unproductive, unnecessary and make the service less attractive, uncompetitive and unacceptable to patients, relatives, and referring agencies. The effort to make services *right first time* is less costly than carrying out a process, eg taking an X-ray, postcoding, blood test, two or three times.

In simplified terms, quality costing includes the documentation and periodic analysis of excess costs arising from having to do operations and tests more than once because of poor care and service, repeat service, or rectifying predictable problems. Identification of high cost areas (10 – 25 per cent of total) which form between 60 – 80 per cent of quality costs, lead to systematic elimination of these areas via quality improvement.

Table 2.1. Values associated with main quality areas

1.	**Mission and core values**	Explicit statements exist within the unit about strategic and operational direction and purpose. They are widely communicated and understood by staff. Objectives for the unit are based on effectiveness, efficiency and quality within the context of business planning and resource management. Meeting customer requirements is inextricably linked to performance criteria.
2.	**Leadership and management style**	Managers continually motivate their staff to ensure high quality service. Staff are treated with respect, given appropriate skills, knowledge and encouraged to contribute new ideas. Traditional entrenched attitudes/barriers between staff groups are reduced or eliminated. Managements' overall style is perceived to be firm and fair.
3.	**Customer requirements**	Regular effective communication exists between *suppliers* and *customers* (internal and external) to refine and improve understanding of changing expectations and needs. Information flow to and from patients on what is expected of patients and of staff services is good. Standards and targets are regularly adjusted.
4.	**Resource management**	Clinicians and managers are both involved in *managing* the unit's resources. Issues of quantity, quality and cost of service are seen as complementary not oppositional. Clinicians, nurses and managers work together at specialty and/or ward level to ensure effective use of resources.
5.	**Standard setting and audit**	Standards exist for all departments and are understood by all staff — zero deviation from these standards is the goal. Capability and performance of processes are subject to regular review. Clinical groups are actively committed to investigating potential changes in practice and learning from *mistakes*.
6.	**Human resource management**	The unit recruits and retains appropriate staff of high calibre to achieve agreed level of standards. The unit trains and educates staff to fulfil determined standards and encourage commitment. The unit promotes participation of staff, through consultation and involvement and recognises the value of staff's views. Staff are encouraged to stay with the unit.
7.	**Communications**	The unit has an effective communications system and regularly elicits staff views on service issues. These views are acted upon. Staff consultative mechanisms (formal and informal) are in operation. Customer forums are held with internal customers (staff) and external customers: patients, relatives, GPs/FHSA, CHC. Positive public press releases are produced regularly based on the unit's quality achievements. Shortening of lines of communication and clarifying accountability at all levels is emphasised in the unit. Communication and working relationships based on trust and integrity are encouraged by the unit. Potential impact of reducing these costs on the quality of service is analysed.

What are quality costs?

Quality related costs can be identified in:

> Strategic planning and implementation of new services.
> Operation of delivery of care and service.
> Maintenance of quality management systems and consequent costs.

So-called *system failures* which incur cost can result in:

> Delay in care being delivered
> Unnecessary testing and treatment
> Repeat care to rectify errors
> Poor patient care and service
> Non-conformance to explicit or implicit standards
> Higher unit costs
> Unnecessary and expensive litigation
> Unnecessary but justifiable complaints and costly, timely complaints' investigations and administration
> Recall of patients for repeat tests and treatment

and **most importantly**
> loss of patient goodwill

It is estimated that fewer than 10 per cent of provider units know how big their quality costs are. This is surely not a good position for trusts and DMUs to be in from April 1991 onwards.

Successful quality costing projects have been carried out by British Aerospace and British Airways, textiles industries, manufacturing and computer industries, the car industry and banking (Department of Trade and Industry, (1989)).

To establish a quality costing system, the following checklist of quality cost components is a useful starting point. In BS6143 the following cost categories are utilised:

Prevention costs	incurred trying to keep failure costs and appraisal costs to a minimum. Made up of money spent or quality management and quality improvement, eg quality assurance staff salaries; quality assurance training costs.
Appraisal costs	correspond to quality control activities: inspecting, testing, auditing and so on, eg supervising staff/managers' salaries.
Internal failure costs	associated with rectifying care and service not appropriately or incorrectly done, using additional resources.
External failure costs	relating to litigation claims and loss of *customer* goodwill; costs of readmissions; continuing dependence on the community services.

Preliminary NHS Quality Costing projects in the NHS identified many areas to which significant quality costs could be attributed. They were:

Area	**Effect**	**Costs**
Outpatient appointment *failures* leading to Did Not Attend (DNA's)	Wasted time of doctors and nurses	Additional clinics
Insufficient car parking	Disrupted clinics: wasted time	
Poor communications with kitchen or meal requirements	Nurse trips to kitchens to collect meals	Increased nursing costs

Slow response to maintenance requests	Time wasted chasing up Estates Department	Increased nursing costs
Incorrect pathology test specification	Waste of pathology service	10 per cent waste
Delay in X-ray reporting	Treatment delays Increased LOS	Increased inpatient costs
Poor radiology appointments system	Poor staff utilisation	Increased radiology costs
Unavailability of porters to take patients to theatre	Theatre sessions delayed; capacity reduced	Increased operating costs
Delayed discharge due to To Take Out (TTO) drugs being unavailable	Increased LOS	Increased inpatient costs
Health records missing	Delays and disruptions in clinics and wards	Wasted nursing and medical time

When the total estimated costs of non-conformance and quality costs are assessed, they total approximately about 10 per cent of the total activity for the service. It is likely that **total** non-conformance costs are about twice the total of these major items which are easily identifiable: 20 per cent!

Having established an appropriate context for costing quality within a hospital or community service, the next steps would be to:

> Establish quality improvement teams in the main area of non-conformance to investigate problems in detail, develop and implement improvements.
> Extend quality costing exercises to ward and department level.
> Develop quality costing information or systems based on simple data collection through surveys and audits.

At ward level, for example, medical and nursing staff would be required to keep records of their daily activities as follows:

Medical staff	Details of delays and disruption experienced; known causes of delays; their effects; time and materials involved.
Sisters	Activities and time spent under headings of: 1. Nursing 2. Health records 3. Administration and computer activity 4. Supervising and training 5. Checking and inspecting nurses work

Associated quality costing information systems should also be developed. These would include:

> Definition of indicators to be monitored regularly and which predict quality costs;
> Use of automatic calculation of costs;
> Illustration of both basic cost data and costs as percentage of activity costs;
> Trend analysis on quality costs.

Why measure health care quality costs?

Quality costing allows quality related activities to be expressed in language which management understands. Although senior managers are now slowly becoming more aware of the broader aspects of TQM, they understand and relate better to outcomes and processes which can have financial implications identified. Bringing quality costs into the general management/business planning/resource management arena helps to emphasise the importance of care and service quality to the overall efficient and effective delivery of affordable health care.

Quality costing, and in particular the measurement of quality costs, focuses management attention on areas of high expenditure and helps to identify potential cost-reduction opportunities. Measurement of quality costs is the first step towards control.

It allows the use of accounting techniques for evaluating the cost benefits of expenditure in certain areas of the service, helping managers decide how, when, and where to invest in preventative activities and equipment.

This process can be used to:

Monitor performance
Identify services for investigation
Set cost-reduction targets
Measure progress towards targets
Cost benefit of individual quality activities
Compare performance between similar departments in different hospitals
Initiate improvements
Uncovering quality problems.

The costs of quality should gradually decrease over the first five years of a TQM programme. This is illustrated in Figure 2.2 which shows that appraisal and failure costs should decrease, whereas prevention costs will increase, giving an overall reduction in quality costs.

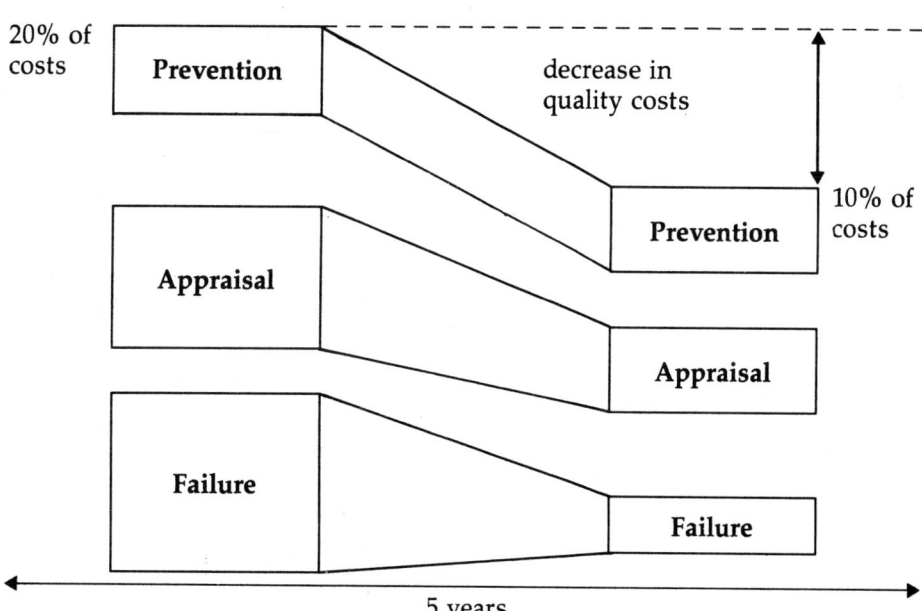

Figure 2.2. Alteration of quality costs via TQM

A crucial factor in continuous quality improvement processes is the collection, analysis and use of quality-related cost information. Provider unit staff will be surprised to find potential savings to be made and the advantages of quality related costing systems for gaining greater benefits and cost control.

Audit instruments for TQM

TQM audit instrument I

The following approach was evolved as a preliminary idea for evaluation, where any hospital or community service is currently developing a total quality service.

The approach consists of a structured interview, covering eight main areas of managing health care:

1. Service provision
2. Service development
3. Finance and manpower control
4. Income generation and releasing resources
5. Organisational development
6. Internal and external relations
7. Estate management
8. Human resource management

Assessment occurs not only of the current status of these eight capacities, but also the extent to which systems are in place to **monitor** and **improve** the quality of these management functions.

Each function is taken in turn. It is not an exhaustive list but includes all the areas covered in the TQM chart (Figure 1.4) and covers the main TQM strategic areas of management culture, quality standards, customer relations, internal staff relations, and training for quality.

1. Service provision indicators

		Monitoring (Examples)	Improvement (Examples)
1.	*Accessibility*		
	a. Waiting times (outpatient) by speciality	No. of weeks	Plans to reduce to 10–13 weeks
	b. Waiting times (inpatient) by speciality	No. of months	Plans to reduce to 12 months
	c. Waiting time after appointment time in clinic	No. of minutes	Plans to reduce to 30 minutes
	d. Time spent with Doctor on first appointment	No. of minutes	Plans to have minimum time
2.	*Effectiveness*		
	a. Standards set for all services	1. Written measurable standards 2. Numerical monitoring 3. Monitoring and review progress in place	Plans to include all services Plans to establish data base
	b. Existence of clinical audit in all medical, nursing and professions allied to medicine	Existence of meetings	Plans for reports to UMG, DMG, District Medical Audit Committee
	c. Infection rates		
	d. Frequency of new pressure sores		
3.	*Acceptability*		
	a. Frequency of complaints		
	b. Consumer feedback		
	c. Cross-boundary flow out of district (and/or region)		
4.	*Efficiency*		
	a. Length of stay		

Examples of **indicators** for other areas of managing health care are:

2. Service development

1. Service strategy for
   a. Acute services
   b. Priority services
   c. Community services
2. Ownership and understanding of strategy
3. Level of cross-agency involvement

3. **Financial and manpower control**
   1. Clarity of organisational management structure
   2. Level of budgeting and manpower information: type, frequency, destination
   3. Level of expenditure
   4. Level of report/establishment discrepancy
   5. Sickness and absence rates
   6. Budgeting implication of resource usage and poor quality
   7. Devolution of accountancy services
   8. Level of identification of efficiency savings

4. **Income generation and releasing resources**
   1. Awareness of quality costs
      a. Internal failure costs
      b. External failure costs
      c. Appraisal costs
      d. Prevention costs
   2. Current income generation (annual)
   3. Cost improvement achievement

5. **Organisational development**
   1. Existence of clear management structure
   2. Aims, values and objectives at:
      a. Unit
      b. Hospital
      c. Department levels
   3. Ownership & commitment to 2 a.b. and c.
   4. Implementation of individual performance review (IPR)
   5. Existence and cohesion of teams
   6. Involvement of clinicians and ward managers in resource and quality management
   7. Unit plan for TQM and ownership and timescales
   8. Delegation of responsibility for QA/TQM
   9. Integration of different functions
   10. Establishment of TQM systems

6. **Internal and external relations**
   1. Communication with external consumers (patients)
      a. Patient information
      b. Patient feedback
      c. Complaints
      d. User groups
   2. Communications with internal consumers (staff)
      a. Communication techniques (team briefing, in-house newsletters, open forum meetings)
      b. Staff feedback and customer – supplier chain analysis
   3. Communication with senior management and authority
   4. Communication with external agencies

7. **Estate management**
   1. Quality maintenance indicators
   2. Labour management system (LMS) indicators
   3. Space utilisation
   4. Land utilisation

8. **Human Resource Management**
   1. Clear mission statement, values and objectives
   2. Ownership of (1) by staff
   3. Industrial relations and Health & Safety machinery
   4. Fair employment practices
   5. IPR process to ward level
   6. Turnover and sickness rates
   7. Training and continuing education
      a. General
      b. TQM
      c. Customer care
   8. Recruitment and retention policies
   9. Incentives and rewards for staff
   10. Implication for staff of poor and high quality
   11. Facilities for staff.

TQM audit instrument II

An alternative type of instrument is the staff questionnaire consisting of many statements relating to aspects of the service and its values felt to be crucial to the development of TQM. The areas covered include:

Areas		Typical statements	
1.	Mission and core values	1.1	I know what the unit is trying to do
		1.2	I know that the unit's business plan is
		1.3	This unit values its staff
2.	Meeting customer requirements	2.1	This unit is clear about what patients expect of staff
		2.2	Staff are aware of what patients want
		2.3	We are committed to finding out what patients want and expect
3.	Communications	3.1	Information circulates through the unit rapidly and accurately
		3.2	Management regularly seek my views
4.	Management and organisation	4.1	Change is well planned and communicated
		4.2	All staff know to whom they are accountable
		4.3	My training needs are recognised and, where possible, met
		4.4	I have regular and positive appraisal meetings with my boss.

A representative group of staff, or all staff, can be asked to complete a questionnaire covering these areas using some form of rating scale, eg five point rating from *strongly agree* to *strongly disagree*. A simple analysis of replies can illustrate the net percentage of *positive* or *negative* results from staff to indicate a significant perception within the unit in a certain direction or any statement. This can be noted for:

group of statements (eg management and organisation)
group of staff (eg nurses)
particular hospital or specialty

TQM audit instrument III

A key aspect of successful TQM is the development of an effective management team. A very useful checklist has been developed by Mathew (1990) to assess management team effectiveness at any level and is shown below:

Management team effectiveness ALPTEC checklist

		As a management team we do this:				
		VERY WELL	WELL	OK	NOT VERY WELL	NOT WELL AT ALL
		++	+	.	−	− −
Aims	Having an inspiring vision for the organisation	....	....	....	....	....
	Being clear about the management team's aims	....	....	....	....	....
	Setting priorities for the organisation	....	....	....	....	....
Leading	Pulling in the same organisation as each other in the team	....	....	....	....	....
	Helping the whole organisation pull together	....	....	....	....	....
	Showing the organisation that we're an effective management team	....	....	....	....	....
Valuing people	Gaining commitment from our staff	....	....	....	....	....
	Listening to what staff are saying	....	....	....	....	....
	Communicating with staff	....	....	....	....	....
	Helping our staff to work well	....	....	....	....	....

	As a management team we do this:				
	VERY WELL	WELL	OK	NOT VERY WELL	NOT WELL AT ALL
	++	+	·	−	− −
In the management team:					
Listening well to each other	....	....	....	....	....
Collaborating with each other	....	....	....	....	....
Showing appreciation to each other	....	....	....	....	....
Trusting each other	....	....	....	....	....
Managing, not suppressing, conflict in the team	....	....	....	....	....
Raising problems about the way the team is working	....	....	....	....	....
Dealing well with these problems	....	....	....	....	....
Supporting each other	....	....	....	....	....
Encouraging everyone to contribute	....	....	....	....	....
Celebrating our successes	....	....	....	....	....
Negotiating with each other	....	....	....	....	....
Task Setting clear responsibilities for team members	....	....	....	....	....
Planning our team's work	....	....	....	....	....
Matching the organisation workload to its resources	....	....	....	....	....
Remedying the organisation's weaknesses	....	....	....	....	....
Managing the organisation's work activities	....	....	....	....	....
Reviewing the organisation's performance	....	....	....	....	....
Reviewing our own performance as a team	....	....	....	....	....
Making decisions together in the team	....	....	....	....	....
Solving difficult problems as a team	....	....	....	....	....
Having the right information on performance of the organisation	....	....	....	....	....
Environment Scanning the team's environment	....	....	....	....	....
Making staff aware of the importance of customers' views	....	....	....	....	....
Knowing who the management team's *customers* are and what they want	....	....	....	....	....
For the management teams' customers					
Knowing what quality is for them	....	....	....	....	....
Delivering what they want	....	....	....	....	....
Checking they are satisfied	....	....	....	....	....
Dealing positively with criticism of our team's work	....	....	....	....	....
Making good links between our management team and other staff	....	....	....	....	....
Responding well to pressure from outside the team	....	....	....	....	....
Looking to the future	....	....	....	....	....
Encouraging change Encouraging creativity and innovation in the organisation	....	....	....	....	....
Encouraging creativity and innovation in each other in the team	....	....	....	....	....
Accepting it when we've made a mistake in the team	....	....	....	....	....
Actually changing the way we do things in the organisation	....	....	....	....	....
Working creatively together in the team	....	....	....	....	....
Performing well Producing excellent performance from the organisation	....	....	....	....	....
Producing excellent performance from the team	....	....	....	....	....

Are there other major areas of activity that your management team needs to do well?
Please list them below so that we can discuss them.

	++	+	·	−	− −
_____	....	....	....	....	....
_____	....	....	....	....	....
_____	....	....	....	....	....
_____	....	....	....	....	....
_____	....	....	....	....	....
_____	....	....	....	....	....

Source: Mathew, 1990

Chapter 3 Organisation and management of TQM projects

Transforming the culture

Total Quality Management (TQM) is a corporate approach which recognises that customer requirements and health care delivery objectives are inseparable. Current NHS organisational culture is very powerful, it can both limit and improve effective health care delivery. NHS staff think and act in most parts of the UK as *carers* or *support carers* in hospital and/or community services. However, this *caring* culture is perhaps not as proactive and supportive as it might be.

Phrases such as 'that's not my job'; 'that's not the way I do it here'; and 'we've tried that, it didn't work' are commonly heard phrases in parts of our services which are not driving quality onwards and upwards.

Most general managers and senior clinicians would like to tackle the negative aspects of their hospitals or community service culture but feel relatively impotent to do so — TQM provides a vehicle or a framework for transforming the NHS culture into the 1990's world of quality driven health care delivery. TQM is about controlled organisational/provider unit improvement towards total quality — an agreed goal for all staff. This is a cultural shift — without it, the hard work put into implementing specific technical components of QA eg standard setting, will not be sustained.

Creating a Total Quality culture in health care

Culture comprises the following:

- People's values
- People's attitudes
- People's behaviour

If you walk around your own hospital or community unit, you will recognise the cultural norms evident in these places. Some are friendly and welcoming with customer receptiveness being very noticeable. Others are superior and non-receptive eg too busy, ponderous, detailed, easygoing. Although there is no stated policy (usually) to back or promote a particular hospital culture, nevertheless it is there. How can you define what your culture is? Areas can be investigated to establish the type of culture existing:

	Poor unit	TQM unit
Customer involvement	Irrelevant and too difficult	Customer driven
Purpose and goals	Unsure and not discussed	Clear and corporate
Activity	Aimless and uncoordinated	Directed
Inefficiency	Allowed and unavoidable	Not tolerated
Levels of management	Several exist and reinforced by management	Kept to minimum and flattened wherever possible
Teams	Cumbersome and uncoordinated	Reinforced and valued
Managers	Administrators	Leaders
Status	Salary related	Personal worth
Quality activity	Rarely evident	Geared to constant improvements
Staff involvement	Variable and inversely proportional to seniority	Staff feel empowered to improve quality
New ideas	Seldom promoted	Innovation rewarded
Moods of staff	Low, helpless	Excited and *potent*

Making the change to TQM

All provider units are endeavouring in some way to go through a quality cultural change, some using TQM. Those trying to address cultural change seriously are achieving quality improvements a lot faster and a lot more efficiently than others. The key elements in transforming culture and *making the change* are:

- Clarity of purpose, core values and objectives
- Leadership
- Planned change process

Having clarity of purpose or *vision* as to where a particular provider unit is headed *enables* and *empowers* staff to see the direction they and you are going and encourage them to help the unit get there. Can your staff articulate their direction and purpose — senior managers, direct care, and service staff?

Many units have taken the time and effort to discuss, brainstorm and develop the following:

- Mission statement — Succinct definition of the purpose of the unit. Four to five lines describing the aims of the unit.
- Core values — General statements addressing the five to eight key components/adjectives in the mission statement.
- Key objectives — Specification of what needs to be achieved to make the mission and core value evident, explicit and recognisable. These objectives take staff beyond current thinking and practice.

Figure 3.1 illustrates a quality policy for one particular health authority which clearly articulates this mission statement and core values, as preliminary steps towards identifying its key quality improvement objectives. This policy was based on the premise that quality management needed widespread investment of time, effort and financial resources. It was constructed via discussions at unit and sub unit/middle management levels prior to an *agreed* policy being approved by the Authority for implementation.

The change process

A coherently planned change process is the third necessary part of transforming the culture. Several lessons can and have been learnt about the right and wrong ways of implementing a TQM programme via modifying the culture of an organisation such as a hospital(s) or community service. It is essential to:

- Reach *all* staff and involve them in our improvement aims.
- Train everyone in achieving improvement.
- Tailor the training to meet the needs of staff in terms of appropriateness content and scheduling.
- Ensure quality improvements are consistent with each other and roughly going in the same direction.
- Facilitate or overcome predictable professional resistance to change based often on fears of losing a skill, role or status.

Five of the most common steps to accommodate the five main points above are:

- Invest sufficient resources to achieve the desired change both financial and human.
- Involve senior managers and clinicians to show their ownership and, as importantly, their own learning needs. Front line staff eg ward sisters, junior doctor, etc must also be involved. The combination of top-down and bottom-up approaches gives greatest results.

Hastings Health Authority
Investing in quality health care (July 1990)

Introduction

This paper sets out the views of staff in Hastings Health Authority on quality issues. It reflects the need to recognise that quality care is not new. In many ways staff have provided excellent care and service for a long time. It relies on all staff feeling valued, valuing each other, and being committed to delivering quality care and service and investing in improving quality.

In this authority, all staff are concerned with providing the best value of health care within available resources. This inevitably involves trade offs and sensible compromises between:

> Quality of care provided
> Quantity of care provided
> Cost of providing care.

The benefits of *investing in quality* are that, consistency of excellent care can be increased, reducing as appropriate some inefficient variability in our service quality. It reduces errors and poor quality. It reduces the wasted resources in not getting things *right first time*. It enhances our morale and pride in our work and vocation. Most importantly, it gives those we serve (the patient, relative, GP, other agency, each other) what they need and expect as much as is possible.

Quality in a provider unit

Quality has been the business and vocation of **all** staff working in the NHS. It comprises:

> **Effectiveness** relating to whether we achieve what is expected and intended
> **Accessibility** in terms of time kept waiting and geographical distance
> **Acceptability** in how the person we serve sees the care and service received

Investing in providing quality should be based on five main activities, parts of which already occur, parts of which are yet to be initiated. These are:

1. **A process for promoting quality**
   This entails defining the standards expected by staff and management for all services, and measuring performance against the standards locally set. This inevitably reinforces areas of existing excellence. It also indicates areas for improvement in quality, especially those highlighted when new skills, knowledge and research become available. Standard setting, monitoring and review should almost immediately identify, and be linked to, the outcome or effectiveness of each service. Information on quality should be available to managers and should cover technical service; information and communication to patient; and physical environment. Systematic and effective processes for internal audit by each department should be in place.

2. **Putting the person first — internal and external consumerism**
   Providing an acceptable service to the public entails us wanting to know people's expectations of us, and whether we have fulfilled these expectations to the best of our ability and resources. Staff do and should stay in touch and identify with people receiving our service.
   Understanding the quality provided entails liaising well with other professional agencies, eg. General practitioners, local authorities, and FPC and other adjoining health authorities.
   Staff in this Authority also *provide* and *use* services from each other. Good quality care and service is based on our internal user supplier chain working well with open and honest communication being everyone's responsibility.

3. **Medical leadership in quality**
   Medical staff have a key role in managing resources, prioritising where developments should occur, and providing leadership and commitment to quality improvement. In addition to their necessary involvement in predicting future activity and the financial implications of this activity, their leadership role in clinical support teams and/or directorate places them in a pivotal role for auditing the outcome of care and supporting changes in clinical practice either on medical or overall clinical team basis.

4. **Training and researching for quality**
   Investing in quality relies on training, continuing education, and research development being acknowledged as essential to maintaining and improving quality of care. Staff at all levels will continue to need support, financial and time, to update their knowledge and skills. This investment will ensure staff in Hastings Health Authority stay at the forefront of their own areas of expertise. In addition, although staff are highly professional, they will require ongoing training in effective methods of quality management.

5. **Ownership in, and leadership of, quality improvement**
   The most important staff are those in face-to-face contact with patients or each other. This means quality is part of everyone's job! It is therefore important that we all *model* high quality work; recognise each other's efforts; support each other's work; and be part of a *culture* which looks for ways to improve. It is essential that managers at all levels (ward, department, and unit) *lead their staff from the front* and inspire them to enjoy improving their services, even though financial resources are inevitably constrained.

Quality in a purchasing agency

With the advent of a new purchasing district with one Directly Managed Unit (DMU) the respective roles of District and Unit with regard to resource management, in general, and quality improvement, in particular, will change.

The new district will become primarily responsible for determining the *appropriateness* of and *access* to the range of services obtained for the population of Hastings. The type and siting of services and the levels and quality of provision will be determined by district in the contracting process. The assessment of the health needs of people in Hastings will contribute to identifying the range of services required. However, the district will need to have a mechanism for evaluating the acceptability of care provided, via elicitation of *consumer opinion*, links with the public and general practitioners and awareness of *cross boundary flows* of patients to other districts.

A crucial initial task is to establish how quality can be specified within service contracts and will involve clinical input from the provider unit and also external advice (eg, Royal Colleges, Region) for this purpose.

To ensure quality of service in Hastings is validly specified, monitored, and reviewed appropriately, and that improvements in care contracted for are achieved wherever possible and affordable, the new agency will need to maintain good working relationships with its DMU management board. This will ensure maximum sharing of information, non-duplication of effort and mutual ownership of one of the main tenets of the White Paper *Working for Patients* namely, improving quality of care.

Action and implementation

This paper has been drafted by the District Quality Assurance Steering Group, discussed with staff throughout the district, and amended in the light of views received. It therefore reflects the ideas and aspirations of the majority of staff.

The intention is to cascade this paper to all departments and, via the normal management process, operationalise the five main components as appropriate for each department in the hospitals and community services.

This will act as the basis for workshops on Quality Improvements to be held in September with Managers and Heads of Departments.

It is also intended to audit current initiatives on Quality Improvement throughout the district and establish an action plan for improvement on the basis of the findings.

Figure 3.1 Quality policy statement

- Continuous and consistent activity geared to achieving service quality improvement. Matching change — actively geared to generalised and widespread improvement with specific 'all-consuming' projects.
- Concentrate on the majority of staff who are not adamantly opposed to TQM, nor absolutely *sold* on the ideas.
- Facilitate cultural change rather than challenge vehemently the existing system.

In addition to auditing the presence or absence of certain aspects or components of TQM, there has to be an audit of satisfaction of those who deal with the provider unit and the services it offers and its aspirations to achieve total quality — namely staff.

The value of a survey of staff attitudes is central to the TQM process because staff — as internal customers of the unit — have important views about the unit, the way it is organised and managed, and the gaps, for instance, in communication. A unit which commissions an internal or external audit of structured discussions with managers and staff will find itself in possession of results which will invariably change the approach to TQM implementation, placing greater emphasis on certain components and underemphasising others.

Such a quantitative assessment of hospital/community service culture is invaluable in that it:

- Structures staff ideas into useable form;
- Allows perspectives of different parts and levels of the service to be assimilated, including those of senior management(!);
- Empowers staff by valuing their contribution;
- Encourages greater commitment at all levels to the development of a locally-sensitive TQM culture

Although not directly/completely applicable to health care, the result of a service organisations' survey cited by Seddon and Jackson (1990) revealed through analysis of a regular staff survey:

- Quality of service given to customers related directly to the understanding staff had of the objectives of the organisation.
- Staff morale and feelings of dissatisfaction were directly linked to the extent to which they felt involved in matters affecting their work.

A useful and simple design for implementing a comprehensive TQM programme aiming to bring about measureable improvements in the quality of care and service which will be continuous and sustained is illustrated in Figure 3.2.

Raising staff awareness

As part of the communications programme, raising staff awareness is essential. As the TQM culture is developing, and a draft quality strategy or policy for the unit is also developing, there comes a time early on in the TQM programme when it becomes self evident that staff at different levels need the opportunity to discuss TQM concepts and practices. This is important for several reasons:

- Clarifying staff views on *Quality* and TQM and its relevance to patient-centred effective care.
- Learning from each other about current quality initiatives within the unit
- Learning from leaders in the TQM field in other provider units
- Clarifying and modifying attitudes to the above approach which are not conducive to offering high quality service.

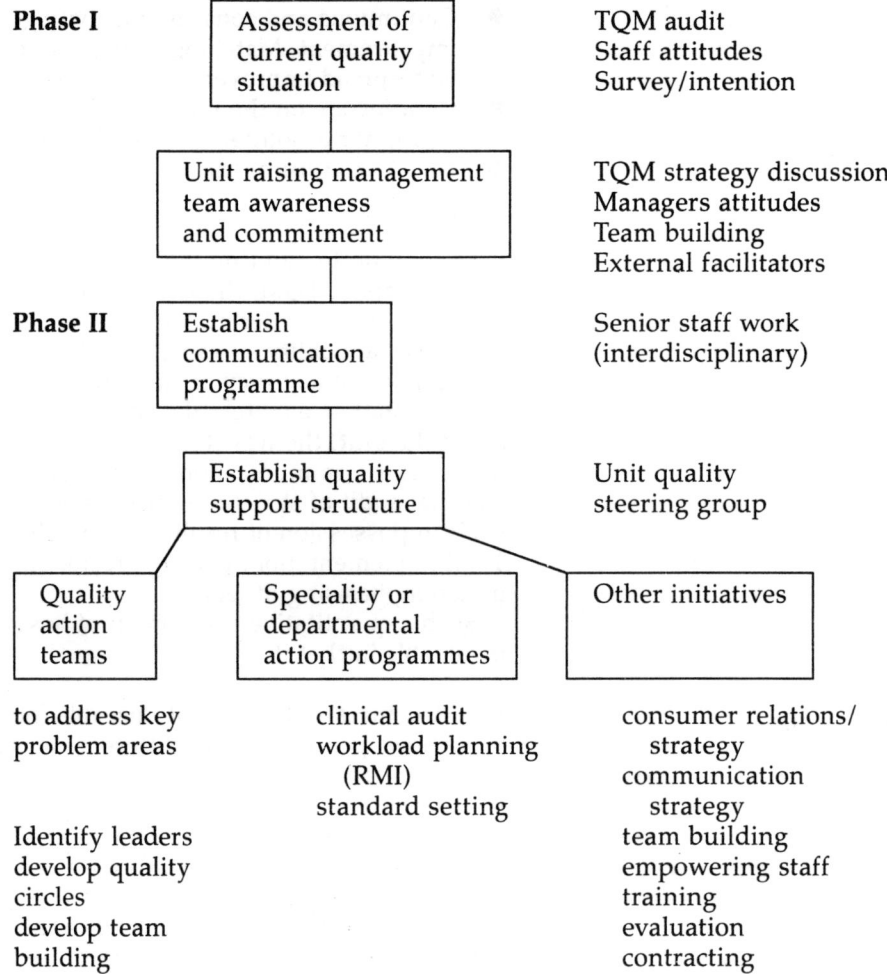

Figure 3.2. Design for implementing a TQM programme

Such staff awareness programmes are relevant to several levels/types of staff, namely:

- Senior Unit Board members
- Consultant medical staff
- Service heads and departmental managers
- Supervisory staff
- Ward Unit managers
- All other staff – clinical and non-clinical.

The key objectives of such programmes should be to:

- Increase staff awareness and understanding of TQM and its component parts;
- Increase commitment via positive attitude formation and reduction of negative attitudes and beliefs;
- Increase identification with the trust of the provider unit management;
- Forge greater working links on quality improvement between clinicians and managers;
- Instil the importance of monitoring quality via the personal approach 'would I be happy with this situation?'
- Encourage staff to be enthusiastically proactive and identifying and implementing service improvements;
- Seek staff views on the developing of TQM programme.

Actioning quality improvements

Having achieved the best possible progress in the following: organisational audit; unit management commitment; raising staff awareness: as a next step it is essential to establish a coherent *structure* for organising both generalised and wide initiatives as well as specific department or specialty programmes. Several key elements have been illustrated in Figure 3.2.

Key coordinating steering group

Aims: to oversee general thrust of quality improvements.
This might be the unit board itself as a specially mandated group which includes senior managers, consultants, plus quality specialists. Usually headed by the Chief Executive/UGM.

Quality action teams

Aims: to address priority areas of improvements on a time limited or continuous basis. Planned outcomes might include:

- Better patient movement services by porters
- Better reception area facilities
- Improved discharge procedures
- Enhanced nursing support services
- Reduction in waiting times in accident/emergency department

If running on a continuous basis they might develop the *quality circle* approach plus address team building issues.

Specialty departments

Action programmes: All departments should be involved as appropriate in drawing up and implementing their own action plans for improvements. The *improvement cycle* should include ways to:

1. Assess current measurable standards and performance against these standards;
2. Identify short falls in performance;
3. Produce action plans to overcome shortfalls with timetables and accountabilities for action;
4. Implement action and measure improvements. Particular approaches would encompass clinical audit, workload Resource Management Initiative (RMI) planning and systematic standard setting.

Other initiatives

Likely to be applicable to all parts of the service:

1. Consumer relation strategy: patient and staff information; customer feedback; internal customer – supplier chain sensitivity
2. Communications strategy
3. Empowering staff
4. Team building
5. Training in quality management
6. Evaluation of change
7. Contracting for quality.

Chapter 4 Information for Total Quality Management

Information, according to a recent Health Minister, is 'the lifeblood of the NHS' (Freeman, 1990). It is information that clinicians use to assess, diagnose, and decide on which treatment to prescribe. It is information that managers require to judge, in close collaboration with clinicians, how resources can best be used and recognise shortfall in resources.

To have the right information in the right place and at the right time is essential and, apparently, in the recent past, has been excruciatingly frustrating to obtain. Information and information technology have a major role to play in the NHS generally and especially in its objective of achieving total quality. To progress explicitly, measurably, towards total quality, provider units will need prime information in all its main component areas:

- Standards and performance against standards
- Consumer feedback (external and internal)
- Costing
- Resource management workload
- Audit
- Communication
- Contract specification
- Coding

Despite viewing actual expenditure on information technology in the NHS with *horror*, comparison with other countries' health care expenditure on information technology shows this to be low, as it is in comparison with other industries eg banking. Increased provision for information technology illustrates this shortfall.

It is important, however, that in the face of huge actual expenditure on information technology, in its broadest form, developments in information in the NHS are justified to produce benefits in improved services for patients.

The proposals in *Framework for Information Systems Working Paper II* (HMSO, 1990) attempted to relate information requirements to the overall objectives of the NHS. Referring to the preliminary action list in this document, the conclusions as to the important information requirements for provider units, to meet April 1991 contracting requirements, are:

1. Improve capture of: a. Postcode
   b. GP
   c. Diagnosis/operation coding
   d. Speciality and departmental cost data
2. Modify Patient Administration Service (PAS) and inpatient information systems to fit with invoicing/contract requirements.
3. Modify/implement outpatient information systems.
4. Additional financial information for contracts: payer identification, price-setting, capital assets, payroll systems.

In many provider units, the coding of diagnoses, operations and procedures remains problematic as well as of variable quality, accuracy and timeliness. The standard of coding will need to improve because of its essential nature for resource management, medical audit and service payment. Equally this applies to postcoding, with its effect on ability to accurately negotiate cross-boundary flow finance.

Resource management, although not an information technology initiative, requires a sound information base to succeed. Considerable effort and expense is being made to implement increasingly complex and integrated computer systems in order to provide higher quality information for decision making by managers, clinicians, and contractors both providers and purchasers. Each major hospital department will have a comprehensive information strategy implemented using the appropriate technology, to support the department's *quality* operation. For example:

> Inpatient admissions and discharges
> Outpatient appointments
> Waiting lists
> Ward ordering
> Theatre scheduling
> Pharmacy
> Pathology
> Radiology

Data can be extracted for purposes such as resource management, medical audit and costing. This would comprise a complete hospital information support system, which links parts of the service as in Figure 4.1.

Ward operations
Diagnostic services
Clinical services
Hotel services
Paramedical services

Figure 4.1. Hospital Services

It can still remain unclear how the multifarious information-related initiatives will dovetail into a coherent practical approach to *understanding* and being informed about care and service to promote better management and improved patient care. Considerable work has been done, for example, in the South Western RHA, to develop management arrangements to *enable* efficient, high quality clinical activity; produce credible clinical data with effective use of information as shown in Figure 4.2.

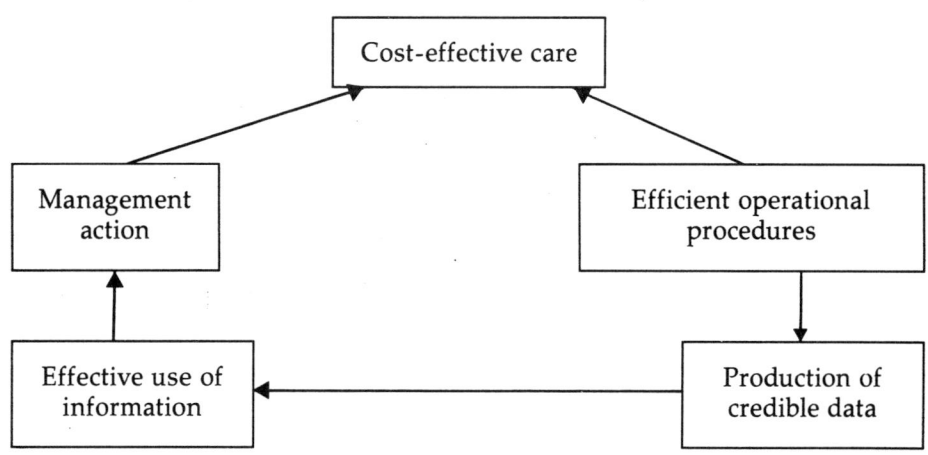

Figure 4.2. Managerial input to improving patient care

Source: South West Regional Health Authority

The managerial input to patient care illustrated in Figure 4.2 is achieved by the effective use of information in planning, delivery, and review of services and the implementation of efficient, high quality operational procedures which *enable* patient care.

Accurate, high quality information contributes to better patient care, in four ways:

- Negotiation of contracts to purchase/provide services.
- Review of use of resources against contracts.
- Agreement/liaison of managers and clinicians within a department/hospital as to what *resource* can provide what *quantity* of care at which *quality*.
- Audit/review of process and outcome of care by groups of health professionals.

Each of these processes involves identifying relevant information needs, production of this information, and the development of competences in using the information. A summary model of how the production of quality clinical information can be achieved is shown in Figure 4.3.

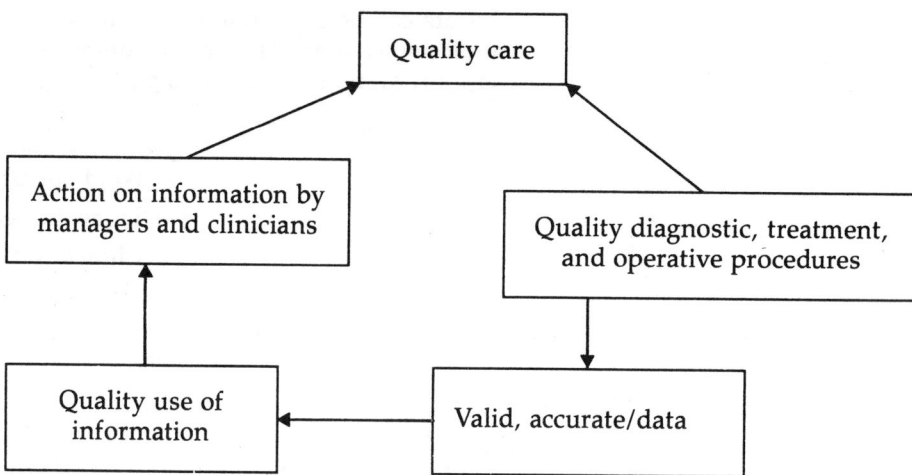

Figure 4.3. Production of quality clinical information

Source: South-West Regional Health Authority

For such a system to work and contribute to TQM, medical staff must be involved, encouraged, and rewarded by taking responsibility for the quality of the data. The capturing and entering of data at the clinical departmental level must be given high priority and the team of coders, secretaries, and junior medical staff involved in data production must be appropriately trained and motivated. Throughout these processes, there must be mechanisms for quality control.

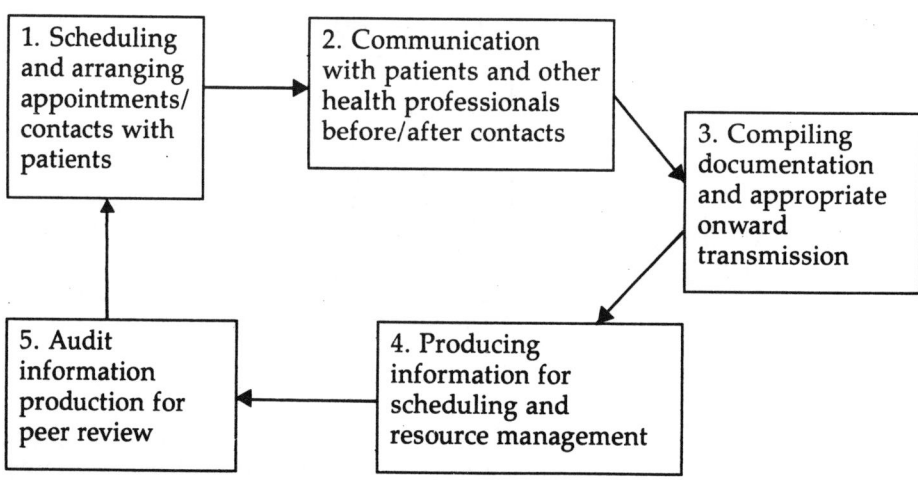

Figure 4.4. Information related tasks for clinical work

Source: South West Regional Health Authority

To carry out high quality clinical work, medical staff require several information-related tasks to be performed, as shown in Figure 4.4. Information production, quality control, organisation of staff involved in information management, should largely occur at clinical unit level. This will result in improvements in data quality, because of the close involvement of medical and nursing staff, close teamwork and a more personal and customer-close approach to patient information production.

As described in the SWRHA approach to *Better Patient Care* and Information Production, *Enabling Clinical Work in the South West* (Mason, 1990), the following tasks can be organised at clinical unit level:

Scheduling and arranging contracts	Making out-patient appointments Maintaining elective admission lists Arranging domiciliary visits Organising the efficient use of beds Arranging emergency admissions within office hours.
Communication	Sending appointments to out-patients Sending for elective admissions Typing and sending letters after out-patient contacts Typing and sending discharge summaries Informing all relevant personnel after a patient contact.
Reception	Receiving patients at out-patient clinics Receiving patients on admission.
Documentation	Raising case notes Ensuring labels are present in case notes Updating contents of case notes Culling the contents of case notes Ensuring case notes are in the right place for patient contacts including pulling them from a library.
Data and information production	Completing management data about patient contacts Abstracting and coding clinical data from case notes Inputting data to the medical data index Completing data about operational performance Producing simple workload and resource use statistics Producing simple clinical statistics for peer review.

All these tasks must be performed to a high quality with monitoring and review processes in place to ensure these crucial tasks continue to be achieved.

Crucial performance standards for measuring key aspects of these patient information processes, which enable clinical work, are as follows:

A. **Health records**
   1. Percentage of duplicate records, reviewed monthly (Target 0-5%).
   2. Case notes pulled/day with no or incorrect location on tracer (or equivalent) system (Target 0-2%).

B. **Inpatient services**
   For all of these the ultimate objective should be no occurrences or 0%.
   1. Routine admissions cancelled by hospital:
      Before admission date
      On day planned for admission.
   2. Routine admissions given less than two weeks' notice.
   3. Routine admissions who failed to attend without giving notice.
   4. Case notes not delivered to ward 24 hours after request.

5. Temporary case notes folders still in use after 24 hours.
6. Computer records containing no discharge date:
    Ward-based systems 24 hours after discharge
    Other systems 72 hours after discharge.
7. Computer records without a current postcode within seven days of contact with the service.
8. Computer records without diagnostic codes completed within four weeks of discharge documentation being available.
9. Discrepancies between the computerised and actual bed states.
10. Discharge letters not sent within seven days of the discharge date.

C. **Out-patient performance criteria**

For all of these, except 1 and 9, the ultimate objective should be no occurrences of 0%. Targets for 1 and 9 should be set locally.
1. Out-patient appointments cancelled by the hospital.
2. Out-patient appointments cancelled more than once by the hospital.
3. Out-patient appointments cancelled by patients because of insufficient notice
4. Cancelled appointments not rebooked within a week.
5. Attendances not on clinic list:
    Fault of computer system
    Attendance initiated by other staff
    Clerical error.
6. Case notes missing from clinic:
    At start
    At end.
7. Temporary case notes folders in clinic.
8. Computer records without an outcome recorded.
9. Patients who failed to attend for appointment (DNAs).
10. GP letters not sent within a week dictation.
11. Appointment not sent within one week of receipt of letter from GP.

Regular surveys should also be carried out to ascertain:
    Patients waiting more than 45 minutes to be seen
    Computer records not checked on arrival
    Investigations missing from case notes when patient seen.

The overall *Case-mix Management* or *Patient Activity Database* (PAD), the integration of many individual systems will require its own systems quality control to apply to systems such as Patient Administration System (PAS), pathology, radiology, maternity, theatre.

The *quality* brief of systems managers should be to support the function of systems by ensuring:

1. Efficient use of systems in support of patient care
2. Integration of systems to support a patient related comprehensive database
3. Proper systems of data protection and security are in place

Tasks are as follows:

1. Formulation of operational policies for the system
2. Development of procedures
3. Responsibility for ensuring that all staff involved are appropriately trained
4. Coordination of the introduction of new policies and procedures
5. Periodic review of existing policies and procedures
6. Setting performance targets and data standards
7. Auditing and monitoring the work of staff involved to ensure that:

Policies and procedures are complied with
Staff training has been carried out
Statistics produced meet the required standards
Performance targets are met
8. Hold responsibility for data protection for that system

Quality and performance indicators

It is often found in units that there is a *split* between professionals interested in *Quality* and managers interested in performance, cost, and activity. This split may be real, but, clearly, in resource management terms and TQM terms makes little logical sense and is wasteful of effort and resources.

As part, therefore, of the diagnostic phase for TQM, time should be spent on establishing how a unit stands in terms of the more common performance indicators that are usually held to assess how *well* a unit is delivering health care:

A. **Service provision**
Activity levels: actual vs targeted performance
Problem factors by speciality
Target-setting and achievement for special areas (eg hip operations, cataracts)
Waiting times and lists size
Performance of waiting list initiatives
Workload plans by speciality
Efficiency of theatres
Surgical reattendance rates
Cervical cytology rates
Bed management and utilisation
Outpatient booking systems
Major review procedures for:
 a. Health and safety
 b. Control of infection
 c. Major accident procedure
 d. Hazard notices
Unpredicted deaths in hospital (Medical and Psychiatric)

B. **Service development**
Sound strategic plans by speciality
Rationalisation of bed distribution
Private sector interface
Implications for services of national reports (eg Confidential Enquiry into Perioperative Deaths CEPOD, National Audit Office NAO)
Business planning by speciality
Service specification and contract development

C. **Manpower and financial controls**
Current budgetary position and recent trends
Associated financial problems
Ability to meet revenue, capital and manpower targets
Elimination of predictable or known overspending
Implementation of cost improvement programmes
Level of cross-boundary flow and need for planning agreements with clinicians to reduce this
Basic management information and its interpretation
Resource management plans

D. **Organisational (unit) development**
Implementation of general management
Management development strategy, and plans
Existence and maintenance of IPR systems

E. **Internal and external relationship development**
Commitment to unit values and objectives
Staff Public Relations (PR) Policies
Unit communications policy
Complaints procedures and analysis
PR policies with external agencies

F. **Management of equipment, buildings and land**
  DGH capital development monitoring
  Use and reuse of vacated space
  General adequacy of space utilisation
  Basic estates database and control plan
  Review of medical equipment replacement programme and procedures
  Energy conservation strategy

G. **Human resource management**
  Manpower control strategy especially nursing and medical staff groups
  Training strategies for major groups and services (eg, psychiatric, Project 2000)
  Skill mix studies
  Nurse productivity and utilisation
  Medical manpower utilisation
  Morale and motivation
  Sickness and absence reduction strategies.

Chapter 5 Standard-setting — ensuring we get it *right first time*

At the time this text is first published only a small number of unit general managers or trust chief executives will have access to complete documentation covering even a small number five – ten of explicit measurable standards for each of his/her departments/wards. In the majority of units, standards-setting will have been variable, inconsistent, and patchy: some services (eg nursing) with comprehensive standards — other services with no written standards. Of those small number of units with comprehensive standards, some may have a process of monitoring and review of the performance of departments against those explicit standards, a process which is *owned* by the direct service deliverers.

At a time when there are at least five different groups of people interested in how quality of care and service is defined and monitored:

Patients
Direct care/service staff (provider)
Unit management (provider)
Purchasing/commissioning board (purchaser)
GPs (purchaser)

There are still very few well defined sets of standards in health care, locally or nationally. Relative to other industries, there has been a reluctance by staff and managers to explicitly state standards by which performance could be assessed and criticised. This reluctance relates to either an intellectual or experiential difficulty with how to make explicit the *behaviour of a lifetime* or the emotional diffidence of the 'highly-trained professional' to others showing 'interest' in how they monitor their own standards.

Before entering into the *bottomless pit* of methodology and possible framework for establishing standards, time should be given to analysing and defining the subject for standard-setting.

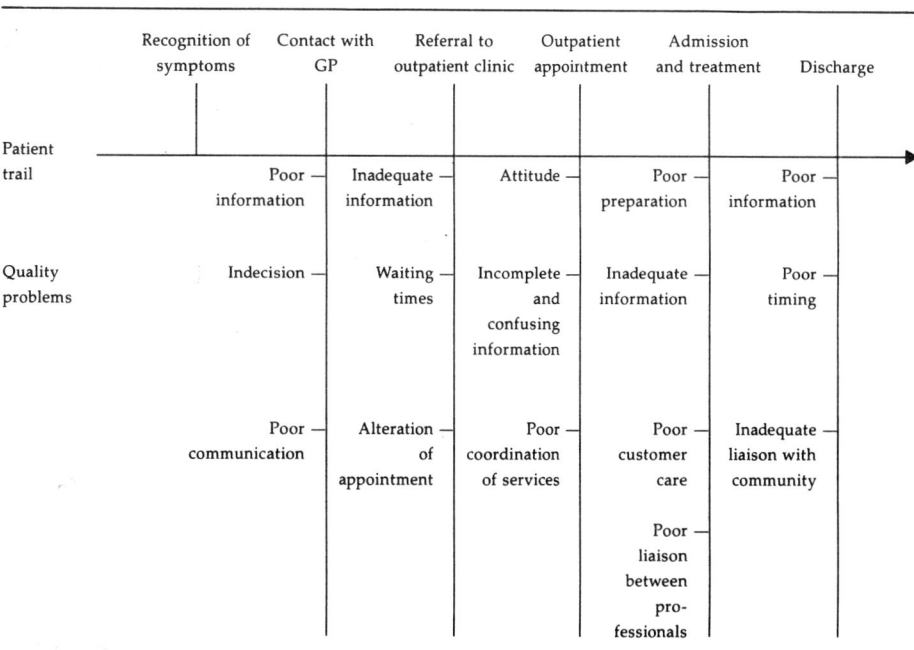

Figure 5.1. Patient trail and attendant quality problems

Source: Hugh Koch

Standards should exist to define:

Care process
For most patients, the *patient trail* model is appropriate with the patient first recognising symptoms and typically seeking help through to diagnosis, treatment and discharge from treatment; Figure 5.1 illustrates a typical trail and typical quality issues along the way.

The quality of this whole process needs to be considered when standards are being set.

Standards for each department/service
The majority of units in which concerted effort has been made to establish measurable behavioural standards have concentrated on standards for each department or speciality. This is a crucial part of the standard-setting process as most staff are trained as particular types of professional with specific skills and competences learnt to a high degree of excellence.

Standards for each health care business function
For those whose sphere of influence involves a relatively high degree of management, if not general management, then standard-setting will need to encompass the main eight management functions of:

- Service provision (quality, volume, range, cost)
- Service development (service direction and vision)
- Financial and manpower control
- Income generation and releasing resources
- Organisational development
- Internal and external relationship development
- Estate management (equipment, buildings, land)
- Human resource management

It is interesting to note that apart from IPR Objectives, general managers in the NHS themselves have no explicit standards, although there is a developing knowledge base for such standards

Standards of outcome, and contributory structure and process
In many services, standards have been defined using the Donabedian model of Structure – Process – Outcome. This has afforded staff a simple model to establish written and measurable standards within.

Structure	Includes the resources put into a service, eg available policies and procedures; numbers of staff; level of training; equipment.
Process	Includes the way in which a service is provided eg accuracy; timeliness; skill in applying diagnostic and therapeutic interventions; information and communication processes; adherence to policies and procedures.
Outcome	Includes ways in which the manager, staff or patient can evaluate whether the service has been effective eg correctness of diagnosis; improvement in patients' health status or symptom remission; level of cross-infection; accidents to patients or staff. This is more difficult to define than structure and process standards and is arguably the most important of the three.

Standards of technical and expressive care/service
Traditionally, providers of health care have often viewed quality as two-dimensional:

Technical service/care — Includes, diagnosis, treatment and aftercare of patients; professional expertise in any department.

Expressive care — Non-technical aspects of care and service to *customers* ie provision of quality accurate and timely **information**; quality **communication** and staff **attitude**; care and service provided in a quality **environment**.

This is perhaps one of the simpler of the models or frameworks for establishing standards.

Standards of strategic vision and development
A bit jargonized! What this means is that the direction in which a service is developing can also predict or form the basis for the framework for establishing standards. For example, in the mental health/psychiatric services, a well planned and resourced *community care* programme will include the following elements, each of which could be addressed from a standard-setting standpoint:

Retained hospital — Most, if not all, community mental health services will retain some formal inpatient services (psychiatric hospital or DGH). Standards can be set within the hospital unit by department or function: acute adult; rehabilitation; elderly.

Rundown of asylum — As the *asylum* hospital is reduced in size, coincident with community unit development, a key quality issue is how to maintain standards on the reduced number of wards with less appropriate patient mix and staff mix.

Community units — Standards of service and care in new smaller community units should be, or aim to be, higher than those existing in either of the above settings. The environment; access to community; and optimism of patients and staff should predict raised standards.

Community mental health teams — Standards for multidisciplinary team work will include quality of liaison; record-keeping; patient case conferences; communication with and access by patients and relatives.

Community liaison — One of the main reasons underpinning community care was the need for higher standards of cross-agency liaison between health, social services, and voluntary agencies.

This model is equally applicable to any area of health care from DGH developments with closure of peripheral hospitals to relocation of patients with learning difficulties.

Standards for consumers?
If a patient was asked to define quality of care or service in any NHS facility, he/she would probably voice three main criteria for high quality:

Accessible — Care should be available as quickly as possible, and the NHS now has targets for outpatient waiting times (12 weeks for first appointment)

	and inpatient admission time (less than 12 months) and as near as possible geographically to the patient's home.
Effective	Care should achieve the reduction of symptoms and enhance the patients' health to the greatest degree given the type of illness and likely course/progress.
Acceptable	Care and service should be offered in the most acceptable way to the patient, in terms of provision of information; receptivity to complaint and feedback; communication style; and staff attitude and environment of care.

These three dimensions, first articulated by Maxwell (1984) could also perhaps be supplemented by his other three dimensions of: **relevance, equity** and **efficiency.**

These seven different approaches to identify standards of health care are illustrated, diagramatically, in Figure 5.2.

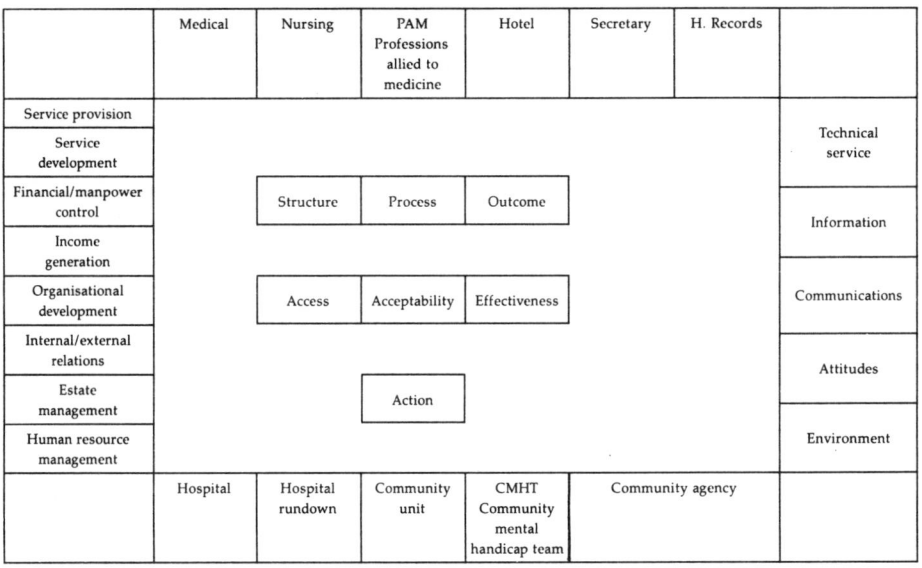

Figure 5.2. Whole care process

Standard-setting can and should be multi-level. Prior to *Working for Patients* legislation, a typical district health authority would approach standard-setting with at least four different levels, as illustrated in Figure 5.3.

At district level, the authority's **mission** would be established and set out as a brief standard statement. This would be backed up by a set of **core values** illustrating the key aspects of the mission's statement. Reinforcing this would be district or unit **policies, procedures,** and general corporate standards. These would be followed by each constituent speciality and/or functional department developing their own standards.

Post-contracting, this multilevel approach becomes even clearer in that the provider-centred approach to standard-setting, which in most districts will be at unit level or below and will need to *fit* with the standard of service expected by its main purchaser.

The purchaser quality specifications will be dealt with in a later section. However, in summary, purchaser-centred standard-setting will emanate increasingly from local networking undertaken by the purchasing agency so that it can be *assured* of the quality of care and service provided by hospital and community unit/trusts with which it lays contracts. An agency will network with several types of consumers-patients/clients; relatives and carers; general practitioners; other statutory agencies eg social services; and

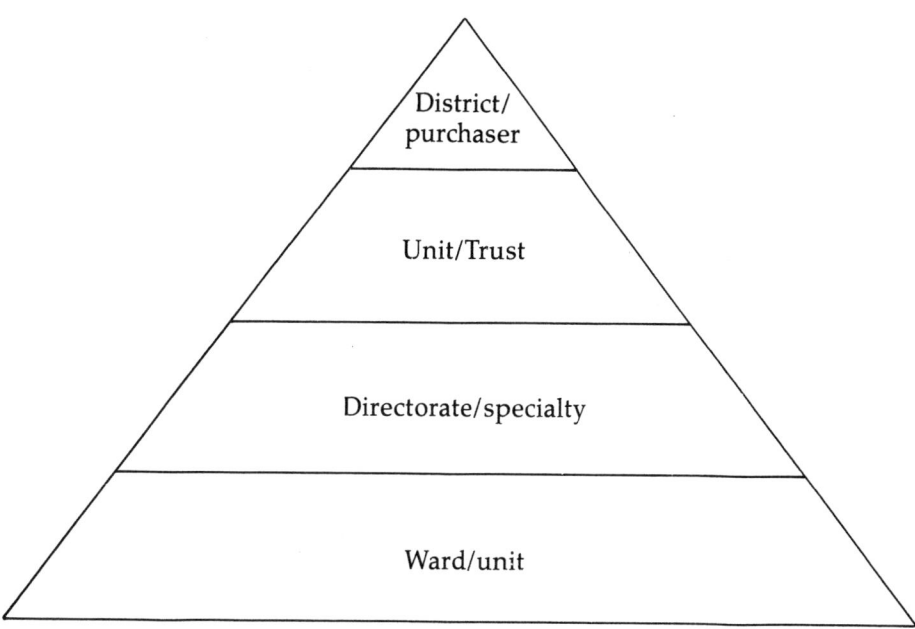

Figure 5.3. Levels of standards

voluntary agencies to establish consumer views and feedback concerning the quality of services currently being purchased.

Before describing the general process of standard-setting in a provider unit, an image which exemplifies the process, is given in Figure 5.4.

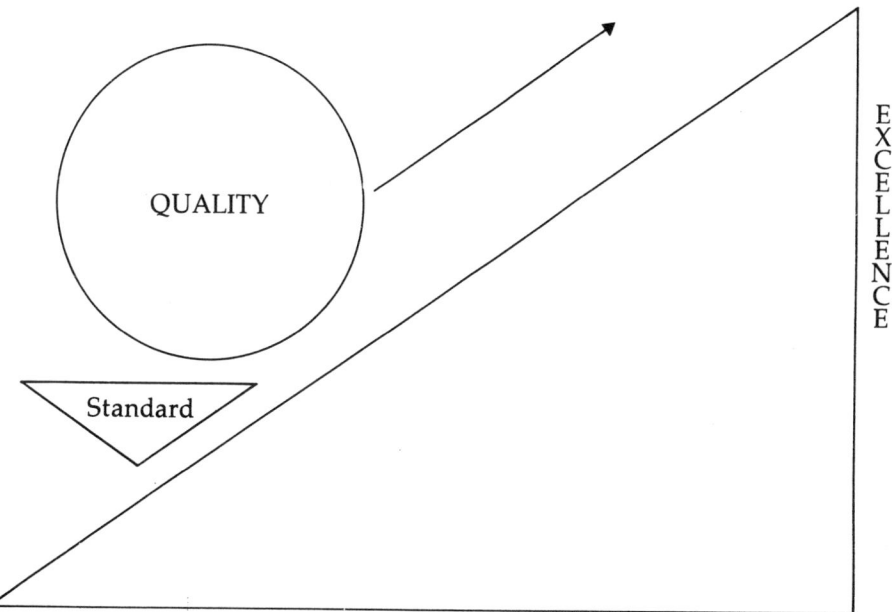

Figure 5.4. Quality 'ball on the hill'

The quality of any service can be represented in Figure 5.4 by the *ball on the hill*. The first issue is:

1. Where is the ball now? What is the current prevailing standard?

To illustrate this image — consider a senior manager's response time to non-urgent mail. The current standard on inspection might be **four days.**

The second issue is:

2. The system should ensure this standard is not reduced.

A process of monitoring/audit/review of this senior manager, is required to put an imaginary *chock* under this *ball* to prevent it rolling **down** the hill. This is not sufficient, clearly, as this prevailing standard may not be acceptable.

The third issue is:

3. Setting a minimal acceptable standard.

To provide a quality service to staff and patients, it may be agreed that senior managers should respond to non-urgent mail in a maximum of three days. The *ball*, therefore, needs to be pushed up the hill and the *chock* replaced at a higher level. In practice the manager may require training, continuing education, support, etc, to achieve this new standard and self or external monitoring to maintain this level.

The fourth issue is:

4. Further quality improvement should always be considered.

The most important aspect of this illustration is the directional arrow pointing up the hill. Services can always be improved: standards can always be considered for improvement. The vertical line, in the illustration, marked *excellence*, does not, in reality, exist. Once a particular *quality ball* reaches or approaches the top of the hill, initially perceived by staff, they will then begin to develop an appreciation of where the hill continues within the landscape. In other words, they will visualize further opportunities for quality improvement. Let's test that with the senior manager example. We might expect that, ultimately, the well-trained, highly motivated senior manager might by now be coming to the office an hour earlier, using the first two hours to clear the daily non-urgent mail, before starting any other business meetings, ie, the standard at the top of the hill is now two **hours** not three or four **days.** Surely, this standard cannot be exceeded? However, the next rise in the hill now appears, as the senior manager decides that instead of dealing with all her correspondence by writing back a letter or memo, she will now deal with 25 per cent by telephoning or dealing with local staff, on a face-to-face level, during that early morning two hour slot, thus providing a much more *personalised management* service.

Participation in standard-setting

There are still many provider units which are starting to develop standards for their services. Some units have progressed a long way with individual services eg nursing, hotel services; others have achieved comprehensive standards across all departments in each hospital or community service. It is variable. With information on content of standards now more accessible, managers face a dilemma: surely it would be simpler, easier, less costly in time, to network nationally or regionally to obtain the best prevailing standards for each service and use these in their own services? This information should certainly be obtained whenever possible, to put a standard-setting process in the context of what has been done elsewhere. However, it is now widely acknowledged that staff benefit from participating in the standard-setting process themselves, to arrive at the set of standards which they *own* and are committed to.

The system for developing standards should support and facilitate initiative and quality improvement. It needs to be well planned and have support both from general managers and key staff at *ground* level. The roles of the quality assurance officer, the general manager, and the *facilitator*, who can develop standards in liaison with the key staff at ward/unit level, are as follows:

General manager — To ensure there is a system for setting, monitoring and reviewing standards; to ensure facilitators are trained and monitored.

Quality Assurance Officer   To develop a framework for setting standards; to support and train facilitators; to maintain an index of unit standards.

Facilitator   To teach colleagues how to write and monitor standards of care; to support and inform colleagues on local TQM initiatives and approaches.

Facilitators are chosen from staff who have expressed interest in TQM and standards-setting and are expected to have training in TQM/standard-setting facilitation. They are very much a resource person for their colleagues.

Training in facilitation of standard-setting involves:

Principles, theory, and practice of setting, writing, and monitoring standards of care and service
Self-directed learning methods
Working with groups
Group dynamics
Leadership styles

Using a *bottom up* approach, so that staff having own standard-setting the facilitators establish a plan to generate standard-setting groups in their own areas. When the standards have been set by facilitators and staff they are *ratified* by the managers (Sale, 1989) to ensure they are achievable within available resources. Dates for review are set for three, six, nine or 12 months' time, when the performance of an area against a set standard is monitored with the continued validity of the standard. This process is summarised in Figure 5.5.

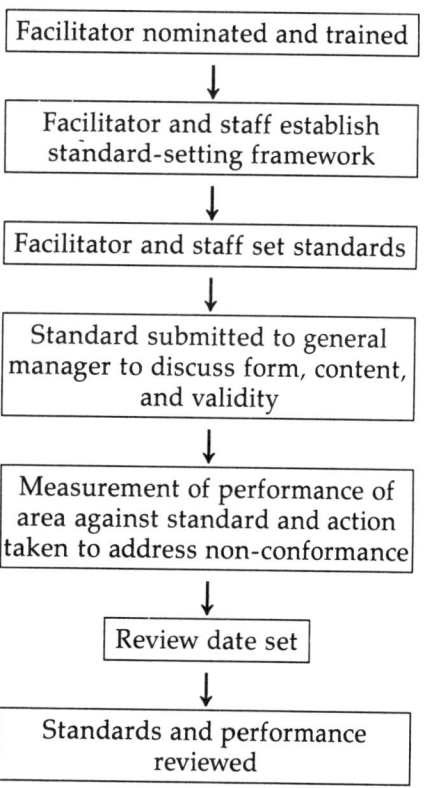

Figure 5.5. *Process for standard-setting*

Standards are not set to overrule or ignore good professional work and judgement, but to acknowledge that performance in any area of care and service can **vary** both **with** individual members of staff over time, and **between** them.

48 Total quality management in health care

Dept/service	Quality indicator	System of monitoring	Standard	Comments
Physiotherapy	1. Length of time on waiting list	Record time on list and review regularly		
	2. Number of DNAs	Keep a record and review individual cases		
	3. Accident and untoward incidents	Record and review regularly		
	4. Outcome of treatment	Compare results of treatment with initial assessments		
	5. Patient satisfaction	(a) Questionnaire (b) Sample interviewing		

Figure 5.6. Physiotherapy standards

Source: East Somerset HA (1988).

Figure 5.7. Surgical nursing standards

Acute unit

Hospital: .... Review: Annual ....

Department: ....Surgical nursing........ Interim ....

Outcomes for good quality service	Achieved	Partially achieved	Not achieved	Further action to be taken
A. Acceptability				
1. Regular clinical training set up with the continuing education department for trained staff	✓			
2. Surgical wounds to be infection-free		✓		
3. Skill mix and staffing levels to be reviewed regularly		✓		
4. Day surgical patients to be received into dedicated day surgical beds		✓	✓	
B. Effectiveness				
1. Care planned during the day to ensure a minimum disturbance for patients at night	✓	✓		
2. Patient comfort maintained and financial costs reduced by implementing a good range of antipressure equipment to meet all needs				
3. Full and correct use of the nursing process documentation with thoughtful and careful evaluation as a means to achieve individual care		✓		
4. Beds to be available with the minimum of waiting time for all booked admissions		✓		

Outcomes for a good quality service Statements of quality assurance	Achieved	Partially achieved	Not achieved	Further action to be taken
1. Structure a. Provision of an area of privacy for treatment				
b. Provision of a quiet and acceptable area for patient treatment				
c. Provision of adequate up-to-date equipment for patient treatment				
d. Regular maintenance programme that keeps equipment in safe working order				

e. Correct OT staffing levels for:
  i. Inpatient treatment
  ii. Outpatient treatment
  iii. Daypatient treatment
  iv. Community patient treatment				

f. Fully staffed OT service in line with current establishment				

Figure 5.8. Occupational therapy service standards (structure)

Figure 5.9.

Standard Reference No:

Topic: Respect

Sub-topic: Privacy

Care group: This standard is applicable in the following wards/departments: Mental handicap unit

Achieve standard by:

Review standard by: Signature of dept manager: _____

Author: Care staff at Copy to senior nurse: _____ Date _____

Standard statement: Each individual has the right to privacy, whilst maintaining the degree of observation required to ensure safety.

Structure	Process	Outcome
1. Private Space		
Using structured games and role play to initiate choice and decision making	To encourage the residents to make decisions on an individual basis	Only enter bedrooms when invited (unless essential)
	Assess the resident's capability to make a decision	To initiate freedom of choice
	Document above to ensure continuity	
2. Bathing		
Enough staff to enable activity to take place without interruptions	Assess individual's ability to bath unaided	Residents are able to have a bath or shower with as much privacy and dignity as possible
Adequate facilities in bathroom (check list)	Document above to ensure continuity	
Lock on door	Wherever possible ensure that male residents are helped to bath by male staff and female residents by female staff	

Source: Cheltenham and District HA

Standard Statement:

....

....

Outcomes/results	Achieved	Partially achieved	Not achieved	Further action to be taken

Figure 5.10. Monitoring and measuring outcomes

This system inevitably involves paperwork and reflects the complexity of most areas of working. How much documentation, with what level of detail, can vary between groups. Key issues are:

- Ownership of standards by all staff in an area
- Identification of non-conformance and action to redress this result from the standards-setting process
- Review and improvement of standards.

The resulting documentation in many units takes the form of a *Standards of Care and Service Manual* providing a comprehensive description of how a service performs; to what standards; and how these are reviewed. It reflects the basic tenet of TQM — that standard-setting should go across all aspects of the unit and should be a *managed* process.

Examples of standards set

Figures 5.6 to 5.10 illustrate several ways in which standards can be set for particular groups. Five different formats are given to reflect standard-setting.

The main criteria for any *output* from standard-setting, is that standards set should be:

- Measurable, ie relatively *behavioural* in their content
- Clear and explicit
- Consistent with service aims and values
- Attainable, within available resources
- Monitored using specified mechanisms.

Monitoring of standards

Having spent a large amount of effort and time in establishing extensive written standards of care and service, it is essential that performance in these areas is monitored against these standards to identify and address non-conformance

The process is summarised in Figure 5.11.

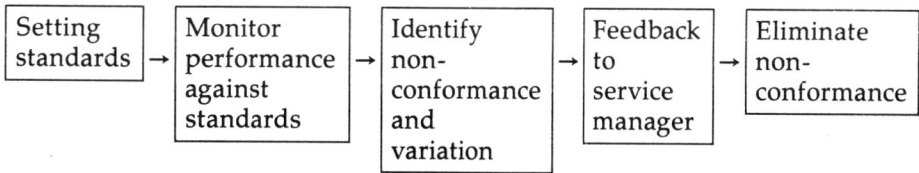

Figure 5.11. Monitoring of standards

Monitoring by staff should be continuous so that key procedures, policies and standards are adequately and consistently controlled. The depth of monitoring will depend upon:

1. Negative/serious consequences of failure or error in service occurring
2. Speed of development of small error/failure to serious damaging consequences
3. Probability/risk of failure/error in a particular part of the service.

An efficient system for monitoring health care and service will be characterised by:

- A number of methods to monitor, deserve and measure
- Involvement of direct care/service staff *and* their managers

When choosing methods of measurement it is important to encompass issues of relative:

validity — what is being measured? It this intentional?
reliability — are the measurements consistent?
sensitivity — can the measurements detect changes that are small but significant?

Given that measurement of performance against standards is worthwhile its relevance and importance of monitoring should not outweigh the accompanying intrusion of the process in disrupting the service. Neither patients nor staff may be unduly inconvenienced. Staff must realise the significance of monitoring and therefore, be committed to accurate recording of information.

The amount of monitoring information given to staff at different levels in the service, needs careful consideration as to its usefulness. The amount should reflect the use and potential exercise of control over a particular part of the service.

Monitoring systems should test whether an appropriate service has been provided to the intended specification in terms of process and outcome. Thus identifying areas of *non-conformance* so that corrective action can be taken.

British Standard BS5750

Many departments and some whole units are considering the application of the British Standards approach to their services quality control.

There is, to date, a paucity of information on the appropriateness and practicalities of this approach within the NHS. Rooney and Wilson (1990) argue for its usefulness in hospital departments such as:

- Outpatient clinics
- Technical services, eg medical physics
- Pharmacy
- Laboratory services
- Radiology departments
- Medical records

The majority of these services have a high technical/equipment orientated emphasis to the service each delivers.

In principle, the components of the BSI approach are face-valid and relevant to any service, including health care. They include components illustrated below:

Policy BS 5750	The decision to use BS 5750 is a major one. Application for BS 5750 Certification is likely to involve major change: a service must be prepared to conform to simple but absolute requirements.
Evaluation	A survey of existing health care systems assesses what changes are needed.
Systems	The systems changes are planned, with detailed design of the new systems (and documents).
Quality Systems Manual	A comprehensive manual with a policy statement signed by the Chief Executive or Unit General Manager and step by step descriptions of how the systems operate is prepared.
Training	Staff are trained in the operation of the new systems and processes.
Operation	The new systems are introduced and run to establish a **track** record of operations.

Internal Audit	The periodic audit of the performance of the quality systems is carried out be staff themselves and verifies that the systems are effective.
Management Review	The periodic review, by management, ensures the continuing achievement of quality standards.
External Assessment	A team of external Assessors examines and report on every aspect of the quality systems with reference to BS 5750.
Registration	The independent Certification that systems comply with BS 5750.
Surveillance	Certification is renewed annually after a satisfactory report by the external Assessors who return during the year.

BS 5750, with considerable success in the manufacturing industries, is gradually finding support in the more *technical* orientated services in the NHS (eg Medical Physics). Some of the key issues in considering the extent to which BS 5750 can be applied throughout the service are:

- Must occur in the context of TQM (or equivalent) ie a *total* process
- Language of BS 5750 needs translation to involve majority of, if not all, NHS staff and services
- Relevance of BS 5750 increases with level of technical/equipment-use services and decreases with social/care emphasis
- Emphasis on documentation needs to be moderated to make more *user friendly*
- It is **one** important method/system for offering a **logical** and **disciplined** approach to quality and **extends** existing processes for documentation (eg medical records, standards setting)
- BS 5750 approach and methodology will *approach* the contracting process.

It is essential that **any** quality control systems *fits* well into a quality culture best typified by TQM, ie a total quality process. The language of BS 5750 requires an immediate translation to fit NHS language — all organisations have their own *jargon*, not least the NHS — without addressing this, the majority of NHS staff will not feel comfortable with labels such as *internal audit, certification* and *registration*. Time spent on translation will maximise the potential applicability of this important approach within the NHS. However, it is likely that the relevance of BS 5750 will be felt most in the more technical services relying mostly on equipment/mechanical services rather than services in which social/caring is predominant. With the overall context of TQM the emphasis in BS 5750 of documentation may need moderating to make it more *user friendly*. Like some other systems for quality control, BS 5750 offers a highly logical and disciplined approach to quality maintenance and improvement and is an extension of existing first level processes for documentation already in existence in the NHS, eg medical records, standards-setting. Finally, it is interesting to note that the new contracting process will require purchasers and providers to adopt a highly structured and explicit approach to quality control, which will inevitably approximate much of what is contained within the British Standards approach.

External standard-setting

In the context of total quality management of health care in the NHS the word *external* can have two important linked contexts, in terms of standard-setting for a *provider* unit.

The first is that the purchasing commissioning agency will develop methods of standard-setting emanating increasingly from its local networking, so that it can be confident of the quality of the services it *buys* under contract from provider units. The commissioning agency will ask a variety of consumers such as patients, carers, relatives, GPs for feedback as to the quality of services.

The second *context* is that of external accreditation — the mechanism for national/regional assessment of organisational standards, systems, and processes for the delivery of health care provided by a particular provider unit. Evaluation of compliance with explicit standards is achieved through survey teams of informed professionals from several health care backgrounds. It is more independent than the present commissioning agency. The King's Fund (King's Fund, 1988) and the South Western Region *Small Hospitals* Accreditation Board (NAHA, 1988) have been leading the way in the UK health care accreditation field.

Potential benefits of external accreditation (King's Fund, 1988) are:

- Development, distribution and updating of national standards
- Identification of *outlier* provider units
- Communication of good practices between services
- Comparison of efficiency of use of resources
- Motivation to confront problems in a wide context
- Peer review facility from outside the provider unit
- Establishment of standards for patient safety, treatment and rights
- Public assurance that good practice exists.

Potential drawbacks are:

- Low participation of provider units in accreditation, unless mandatory
- Standard-setting is beset with problems in terms of uniformity, flexibility, linkage to beneficial patient outcomes
- Lack of provider unit staff involvement.

The process of marrying up the external accreditation process(es) with internal standard-setting for provider units, has already started in both contexts described above. The former is likely to be more successful in the short term as the contracting process is mandatory from April 1991, for all commissioning agencies and provider units, whether they be trusts or Directly Managed Units (DMUs). This process will ensure a much higher profile being given to the existence of standards for all services which are measurable and regularly monitored at both provider and purchaser levels.

Chapter 6 Clinical audit — improving care through audit

A critical part of the NHS agenda has always **implicitly** been to improve quality of care and obtain value-for-money. Since the implementation of General Management and the recent legislation, this has been made more and more **explicit** by introducing systems to review quality of clinical care received by patients. Throughout the NHS, a variety of types of *clinical* audit are being introduced by medical, nursing, and paramedical/PAM professional groups. The main methods and accompanying issues in implementation will be discussed in this chapter. In addition, the gradual development of **Integrated clinical audit** — a multidisciplinary approach to auditing care will be described.

Before describing the variety of approaches being used, the possible benefits of audit are:

- Greater consistency of care — with the quality of care by one or more professions being superlative!
- Raising overall quality of care — moving consistently to higher and improved levels of quality of care
- Reduced clinical errors — errors can occur in any technique (with or without the benefit of *hindsight*) and audit can result in reduced frequency of preventable mistakes
- Improved cost-efficiency — costs of poor quality care can be significantly reduced
- Improved clinician confidence — audit by peer review or alternatives, if carried out sensitively, leads to improved confidence in skills and limitation of care.

The NHS Management Executive *Nursing Care Audit* (1990) outlines the main method of professional self-improvement leading to the enhancement of quality of care delivered to patients. Although addressing **nursing** quality, its basic premises apply to all clinical professions in direct patient contact. It indicates that audit:

- Is a simple process
- Encourages the measurement of performance
- Recognises good practice
- Facilitates improvement.

The cycle of audit activity (NHSME, 1990) which results in a systematic improvement in clinical practice is shown in Figure 6.1.

Audit can be divided into several key principles, which are applicable to any clinical area or professional group:

- Define team/professions objectives — mission statement; core values; key objectives of service
- Develop standards — must be measurable and easily communicated to others; can result in raised consistency of care and evaluation of performance
- Monitor and improve/change standards — a process which is *owned* by the staff group

- Communicate the outcome of auditing — to other members of the same profession; to other clinical professions and managers.

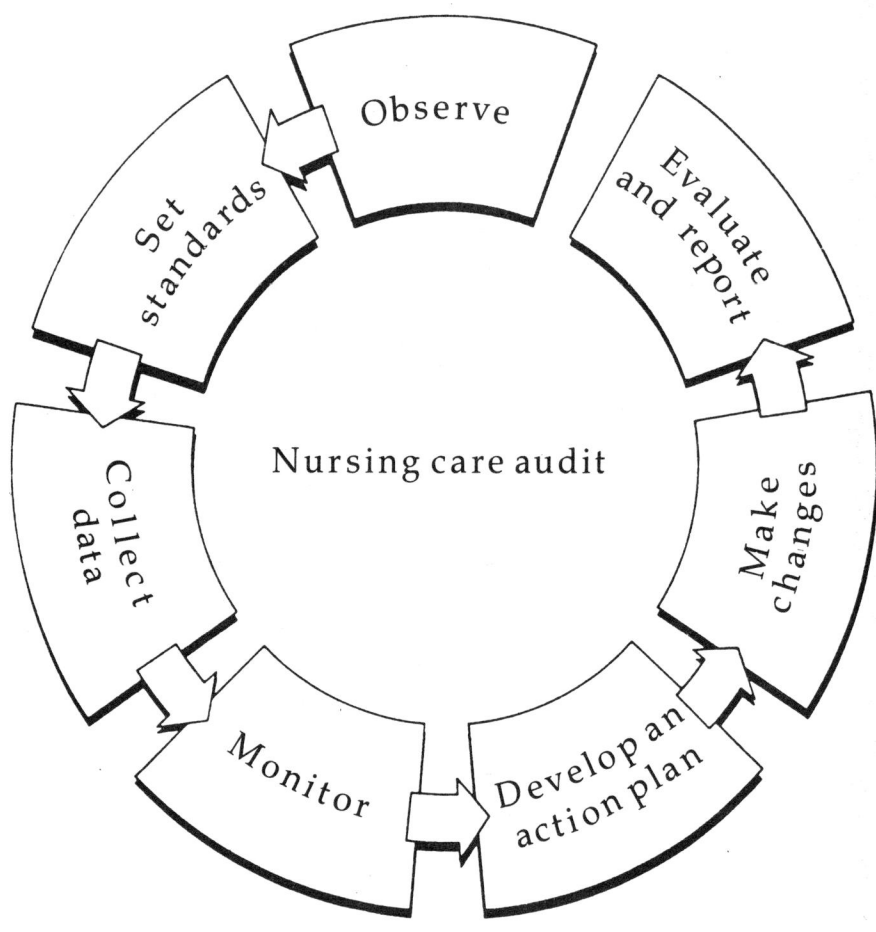

Figure 6.1. The process of audit

The main steps in establishing an audit which is practical, useful and results in improved quality of care in any profession or team are:

- Discuss audit principles and gain support — and commitment of colleagues.
- Identify an area to audit — such as pain management, cross-infection.
- Identify key staff in audit process — define audit group, their objectives, and the necessary time and resources required.
- Develop standards of care to measure current performance
- Collect reliable and valid data — from appropriate sources such as patient records, care plans, complaints, incident reports. Identify new sources of data to be developed.
- Review and discuss results of audit — to identify *non-conformance* with agreed standards and need for specific remedial action.
- Develop *non-conformance reduction* action plan — specifying required action, its measurement, time scale, plus those accountable for implementation.
- Implement action plan

Prior to exploring the practicalities of audit within the main clinical groups, several key issues are worth considering. They were discussed at a successful audit conference *Action on Audit* in London (*Healthcare Independent*, 1990):

- Context of audit

    Audit is defined as the systematic, critical analysis of the quality of care including diagnosis, treatment, resource use and eventual outcomes. It is therefore linked to other key initiatives — **resource management** and **total quality management**. All three initiatives are themselves ways of illustrating and achieving effective management and delivery of health care. It is totally simplistic to caricature each as only covering certain aspects of this mammoth endeavour — **all** are concerned with **quantity, quality** and **cost** of health care.

- Organisation of audit

    Audit must be owned and led by clinical staff. It cannot work if imposed, or felt to be imposed, from management and would fail if this were the case. It is worthy of note that prior to the 1990 legislation, many doctors (and nurses) were already carrying out various types of clinical audit. This then became the norm as a result of *Working Paper 6* in the 1990 NHS Legislation. It is essential that both senior and junior clinical staff are represented in audit arrangements at hospital and department level. Junior staff have much to contribute as well as learn from audit. Audit must be organised in a way which leads to a change of behaviour and practice. *Organisation* includes the issue of time — audit takes a significant proportion of time allocated to clinical work. Consultant medical staff are successfully negotiating a 10 per cent allocation of time to audit. Other professions will no doubt negotiate the same time allocation. The *loss* of clinical time and *quantity of care* will be balanced, in successful audit, by the changes in practice implemented.

- Confidentiality

    Openness and honesty amongst colleagues in audit discussions will only be achieved where there is absolute confidentiality: in the discussions; reports and analysis which staff undertake; and the use made of written reports. It has been suggested that audit meeting documentation which itemises particular clinical case discussions should be kept to a minimum for fear of *sub poena* in law. The conflict between, on the one hand, statutory responsibilities of purchasing agencies, the Department of Health, and lawyers, and on the other hand, responsibilities of clinical staff to be honest with themselves about possible errors and mistakes which, if discussed openly and sensitively, could lead to improvement in care, are considerable. There needs to be a sensible balance between the rights of the individual and the integrity of honest clinical audit.

- Information and information systems

    Reviewing clinical practice must be facilitated by valid and reliable information and supporting systems. Clinical staff must be involved with the production of information and committed to the integrity of this information, which must be patient-centred and clinically relevant. The development in most acute hospitals of the ward/speciality-based Clinical Work Station (CWS) where secretarial and/or health records coding staff input data is crucial to the availability of valid and reliable information. These staff are part of the overall clinical support team.

- Cost and support

    There are several areas of possible cost in establishing effective audit systems and processes:

    1. Clinical time diverted to audit

2. Alternative use of coding clerks and medical secretaries
3. Information technology
4. Training and continuing education
5. Implementation of improved practices

However, taking the five areas as a whole, there is likely to be a cost implication requiring management support, but applying TQM, including charge, costs can be contained.

Medical audit

Lessons have already been learnt from the different types of audit exercises which have been carried out, over the past fifty years of the NHS' existence especially:

1. *Confidential Enquiry into Maternal Deaths*
2. *Confidential Enquiry into Peri-Operative Deaths (CEPOD)*
3. *Perinatal Mortality Surveys*
4. *Surgical Audit*
5. *Alment Committee Report (1976)*
6. *Elderly Discharge Arrangements (Currie et al 1984)*
7. *Asthmatic Deaths (NETRHA 1987)*

In addition, as the public are becoming better informed about medical practice and treatment options they are showing greater interest, as is the Department of Health and other statutory bodies in many key areas of medical intervention. Some examples of this interest are:

1. Radiological procedures in cancer treatment
2. Variation in mastectomy procedures (radical, partial, etc)
3. Caesarean birth rates and criteria
4. Prescribing of psychotropic drugs and ECT.

Traditionally, audit has involved the following elements:

A. **Structure**
   Divisional/speciality structures in hospital medical groupings covering:
   1. Drugs and therapeutics
   2. Infection control
   3. Postgraduate medical education
   4. Medical records
   In addition, aspects of facilities available and levels of staffing are included here.
B. **Process**
   Regular meetings
   Audit reports
   Provision of accessible and acceptable care.
C. **Outcome**
   Documented changes in clinical guidelines and clinical practice
   Changes in avoidable mortality rates
   Consideration of morbidity
   Identification of defined end points
   Effectiveness and acceptability of care to patients.

Examples of areas which can be measured are given in Figure 6.2 taken from Hopkins (1990).

Currently, the nature of medical audit activity typically includes eight areas:

1. Selected case note (or X-ray) analysis
2. Random case note analysis
3. Examination of specific activities
4. Review of the use of investigations
5. Examination of complication rates
6. Examination of outcomes
    general
    deaths
7. Analysis of routinely available data
8. Multicentre activities eg CEPOD.

Structure	Available capital facilities that meet defined standards (hospital accreditation) Number of personnel with appropriate training Appropriate training programmes
Accessibility	Affordability Distance and transport facilities Waiting lists for care

Process	Deviations from normative standards. This requires the preparation of guidelines for good practice based upon:
	    Consensus conferences
	    Meta-analysis of published work
	    Clinical decision making
	    Establishment of efficacy and effectiveness
	    Definitions of appropriate care
	Fragmentation vs continuity of care
	Variations in care
	Imparting information and interpersonal skills
	Economic assessments of effectiveness

Severity of illness	Health status before intervention

Outcome	Mortality, adjusted for case-mix severity of illness
	Morbidity
	    Disease-specific indices of morbidity
	    Measured general health status
	    Perceived health status
	Satisfaction with care

Figure 6.2. Examples of areas in which quality of care can be measured

Source: Hopkins A 1990 *Measuring the quality of medical care* Royal College of Physicians Publications

Analysis of activity is usually categorised into:

- Outpatients
- Inpatients in same speciality (one hospital)
- Inpatients in same speciality (hospitals in one district)
- Inpatients in same speciality (hospitals in region)
- Inpatients in same speciality (UK).

Audit activity is typically mostly, of one-two hours' duration.
Key aspects of medical quality management can be summarised as:

District audit committee
All speciality involvement
Administrative and technical support
Sessional time commitment
Responsibility and accountability at senior and junior levels
Inventory of scope and frequency of existing audit review mechanism
Progress standards setting
Ensure medical staff participation (all grades)
Set up process for quality improvement
Ensure regular audit review.

Audit activity can and should occur at several *levels* within a provider unit.

For example function/issues of audit and quality of medical service at four unit levels might be:

1. Overall quality of medical service — Royal Colleges standards
   Written standards
   Automatic review

2. Speciality audit — Written standards for each speciality
   Annual review

3. Ward/clinical team audit — Policies, procedures, standards

4. Individual patient audit — Assessment of planned activities against actual care given
   Training identification
   Evaluation of care

As was mentioned in an earlier chapter, clinical audit is an essential part of the *Total Quality Management* process, alongside many other key quality processes.

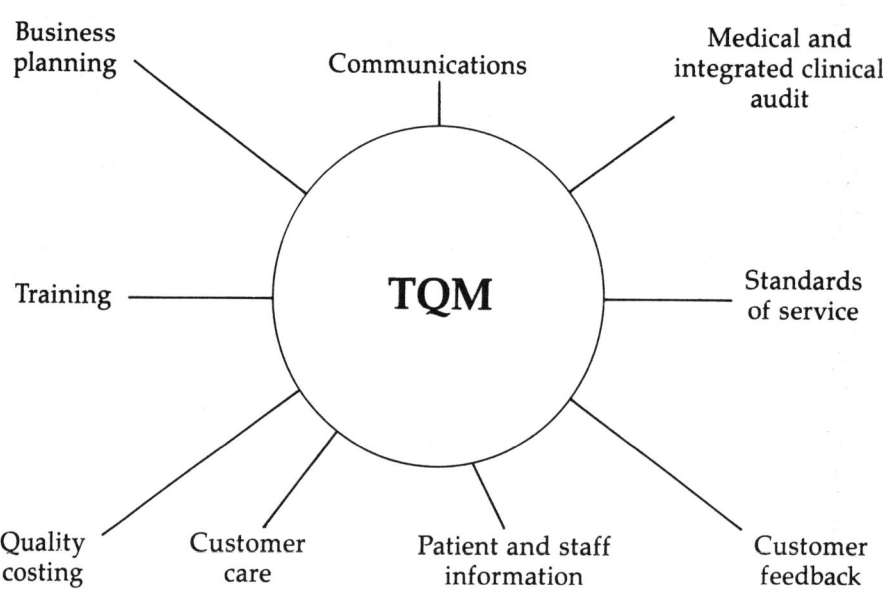

Figure 6.3. Total quality management process

Generally medical staff consider implicitly all the time and explicitly via medical audit the extent to which their services cover Maxwell's (1984) six areas of quality:

1. Accessible
2. Equitable
3. Relevant
4. Efficient
5. Acceptable
6. Effective

These can be considered hospitalwide by all specialities, as illustrated in Figure 6.4.

Clinical audit 63

Quality issues	General surgery	Orthopaedic surgery	ENT	Ophthalmology	Oral surgery	General medicine	Paediatrics	Obstetrics & gynaecology	Radiotherapy	Accident and emergency	Elderly	Mental illness	Mental handicap
Accessibility													
1. Waiting times													
2. Geographical access													
3. Time for relatives													
Acceptability													
1. Communications													
2. Attitudes													
3. Environment care													
Effectiveness													
1. Degree of specialisation													
2. Diagnostic *success*													
3. Treatment *success*													
4. Follow up													

Figure 6.4.
Specialities

In addition, TQM culture and care values need to include the medical view of quality and address the following issues key to the success of quality management:

- Do medical staff put patients and carers **first?**
- Are they fully aware of patients' expectations?
- Do they make every possible effort to satisfy each patient first time?
- Do they recognise and **own** the costs of poor quality?
- Do they support their colleagues?
- Do they encourage staff loyalty, at all levels of the hospital?

As regards patient satisfaction, the use of surveys, questionnaires and structured interviewing utilises questions to elicit patient views of various aspects of care of which those below are typical:

A. **Doctors' overall care**
   1. Was the way in which the doctors cared for you:
      Remarkably friendly — pleasant — mood dependent(!)?
   2. Did you have confidence in the way the doctors treated you?

B. **Information from doctors**
   1. Did the doctors go out of their way to make sure you understood your condition and its treatment?
   2. Did the doctors explain technical language used?

C. **Advice on leaving hospital**
   1. Were you given explicit and, where appropriate, written instructions?
   2. Were your anxieties on leaving hospital acknowledged and received?

General methodology of audit

An elementary starting point is the identification by speciality of:

1. Key clinical problems/symptoms seen
2. Key diagnoses/diagnostic groups
3. Key treatments
4. General quality issues.

This can be illustrated in the form of an *audit grid* which is completed by discussion within each clinical speciality. Examples of the application of this approach are given below for ophthalmology, ENT, general medicine, gynaecology and geriatric medicine.

Moving to more sophisticated approaches, a useful framework for medical audit methods is shown in Figure 6.6 (Offen, 1990) and identifies five main foci of activity in audit.

Typical outcome indicators which have been developed (Hopkins, 1990) are given in Figure 6.7.

The main principles of outcome indicator development (Owens, 1990) are:

1. Keep it simple
2. Be clear on use of indicators
3. Focus outcomes on high risk, high volume, problem-prone aspects of care
4. Rely as much as possible on existing data.

Clinical audit 65

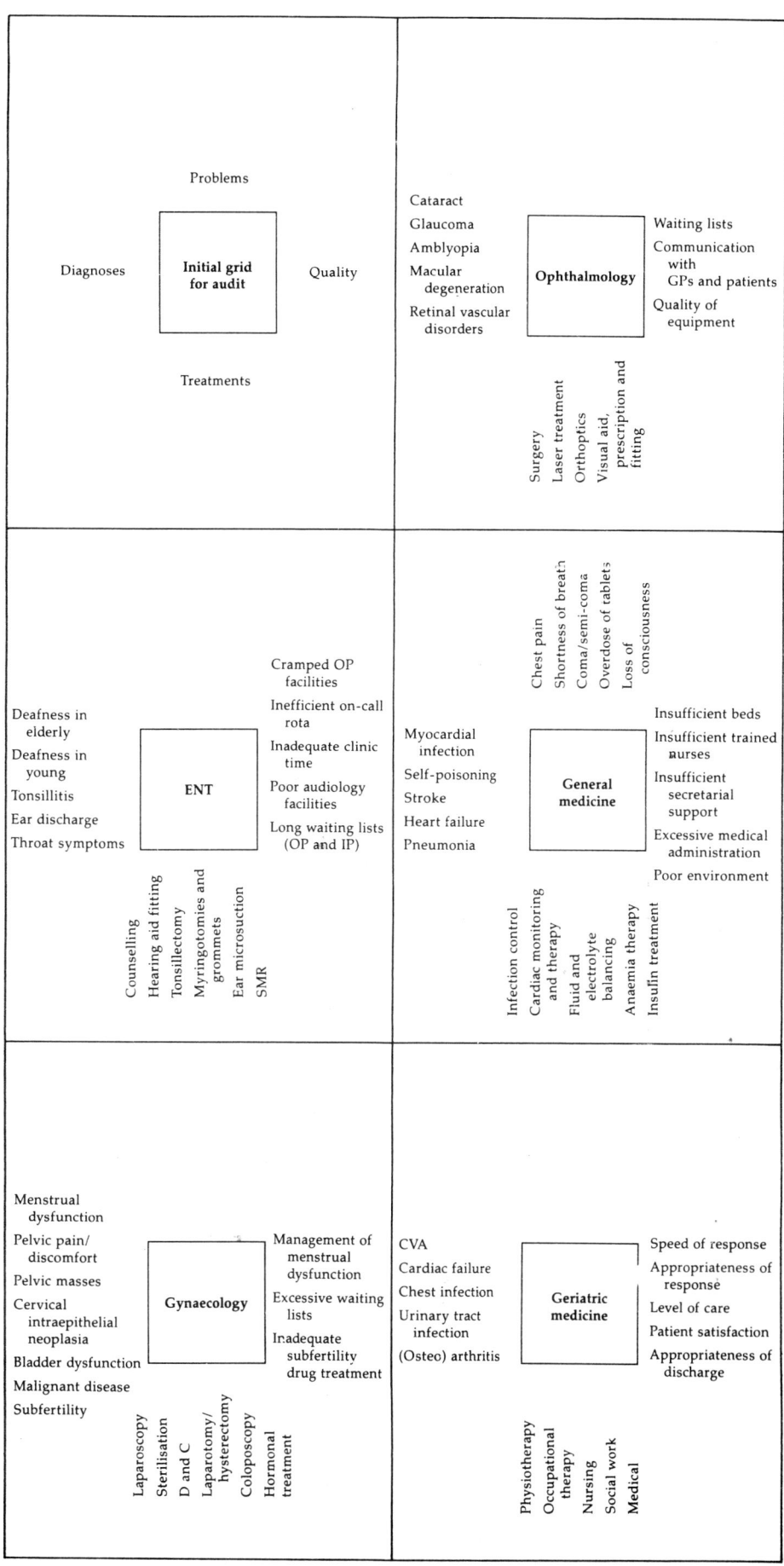

Figure 6.5. Audit grids

66 Total quality management in health care

Focus of activity	Short-term outcome	Long-term outcome	Case Note review	Occurrence screen	Management information
Speciality	Surgical	Surgical Medical	Medical Community based	All	All
Audit items	eg Mortality Morbidity Infection Wound Chest Urinary Pulmonary embolism Deep vein thrombosis	Down load short term outcome Long term sequelae eg: angina, incisional hernia Patient surveys: Satisfaction Residual symptoms Achieved versus expected outcome	Criteria within specific treatment group eg: myocardial infarction: Criteria selected with yes/no answers, analysis, change of policy Criteria re-examined Results *compared* between years	eg Unscheduled return to operating theatre Suicide Bedsore	Length of stay Throughput Turnover interval Waiting list Information Operating theatre usage Readmission rate Cancellation rate

Figure 6.6. Framework for medical audit methods

Source: Offen (1990).

Cardiovascular outcomes	1. Intra- or post-operative cerebrovascular accident (CVA), in patients undergoing isolated CABG procedure	2. Patients undergoing attempted or completed PTCA, in whom lesion attempted is not dilated	
Anaesthetics outcome indicators	1. Mortality within a specified time following anaesthesia	2. Aspiration of gastric contents with development of typical X-ray findings of aspiration pneumonitis, within a specified time following anaesthesia	3. Dental injury during anaesthesia
Obstetric outcome indicators	1. A maternal length of stay of more than 5 days after vaginal delivery or more than 7 days after caesarean section	2. Maternal death rate up to and including 42 days post partum	3. In-hospital intra partum death of the fetus weighing 500 grams or more	4. Perinatal death rate of an infant weighing 500 grams or more
Oncology outcome indicators	1. Survival of patients with primary cancer of the lung, colon, colon-rectum, or female breast by stage, in histological type	2. Patients with primary rectal cancer undergoing abdominal perineal sections, with 8cm or more of free distal surgical margin present on specimen, as documented in surgical pathology close description		
Trauma outcome indicators	1. Return of patients to operating room within 48 hours of completion of initial surgery	2. Death of trauma patients with *one or more* conditions who did not receive surgery for the condition		

Figure 6.7. Examples of outcome indicators

Source: Royal College of Physicians, London, March, 1989.

Occurrence screening (adverse event reporting). Examples

Unplanned discharge
Suicide (or attempted suicide)
Unplanned return to operating theatre
Death as direct result of procedure eg anaesthetic, invasive procedure
Death from nosocomial infection

Management information useful for audit:

Length of stay
Throughput
Turnover interval
Day case rate
Cancellation rate — by unit
— by patient
Operating theatre usage: eg cancelled sessions
timing of sessions
Waiting list — in-patient
— out-patient
DNA rate — in-patients
— out-patients (new and old patients)
Readmission rate (within 30 days)

These main types of audit methods form part of an overall quality specification for a medical service which follows a general template model developed by Offen (1990) and an example of which is given for general surgery:

Template for quality (Offen, 1990)
General surgery
A. **Access**
Out-patients
1. Waiting time for new patients to receive notification of appointment:
Emergency — target two working days
Urgent — target seven working days
Routine — target four weeks.
2. All patients to have individual appointment times.
3. Instructions and guidance for new patients:
What to expect and do on arrival
Who they will see
What will happen
What to do with referral letter
Urine specimen.
standard — 90% seen by consultant or senior doctor.
5. Waiting time for investigations eg contrast radiology
— target seven working days.
6. Old patient case note review by senior doctor before start of clinic *or* senior doctor sees old patients every third visit.
7. New patient/old patient ratio
Standards to be agreed by department.
8. All reports for general practitioners to be written within 48 hours of consultation — Standard: 90%
All reports to be received by general practitioners within seven working days — Standard: 90%.
Management to ensure distribution of reports by first class post or courier service

B. **In-patients**
1. Information for patient
Where to go

What to bring
Who to ask
What to expect
2. Information for next of kin
Where patient is
Visiting arrangements
Who to ask for information
Arrangements for staying with the dangerously ill
Catering and telephone services
3. Waiting time
Urgent within two weeks
Routine within three months
All patients to be given agreed booked admission date at time of decision to admit — target time for achievement to be agreed by department
List of patients able to come in at short notice to be kept
4. Patient cancellation rate
by unit
by patient
Information to be included in management information section of medical audit
5. Timeliness of operation
Operations cancelled within 24 hours of scheduled time
Information to be included in management information section of medical audit
6. Grade of doctor carrying out procedure
Operation
Other
All procedures carried out during normal working hours to be done or directly supervised by consultant or senior doctor except where specifically stated.
All procedures carried out outside normal working hours and lasting for more than two hours to be done or directly supervised by consultant or senior doctor.
7. Ward round by senior doctor at least once weekly.
8. Access to appropriate nursing skills
Also access to: physiotherapy
occupational therapy
speech therapy
support service eg pathology; radiology; nutrition
chaplaincy of appropriate denomination.
9. Access for general practitioners
Unit switchboard
Information for direct access to senior doctor or deputy.
10. Access to day surgery
Policy for selection of patients to be treated as day cases.

C. **Communication**
1. The doctor or nurse dealing with the patient at any one time will keep the patient fully informed of what they are doing and why. The senior doctor and/or senior nurse will inform the patient of the plan for their diagnosis and treatment and of the results as they become available.
2. Discharge reports
Minimum content to include main diagnosis or reason for admission; significant procedures; findings and complications; medications on discharge.
Discharge report to be written within three working days of event — Standard: 90%
3. Death in hospital

General practitioner to be notified within 24 hours usually by telephone
4. Next of kin
Departmental policy for access to consultant, junior doctor, senior nurse. Arrangements to be clearly stated in information leaflet for relatives.

D **Discharge policies**
All patients to be discharged in accordance with unit discharge policy. To be specified by department

E. **Medical audit**
1. Type or types by department
   Short term outcome
   Long term outcome
2. Management information
3. Occurrence screening
4. Infection rate
5. Normal tissue audit
6. Structure of audit committee
7. Reporting activity of audit committee
8. Discharge check and information
9. Pathology standards
   The pathology department participates in the National External Quality Assurance Scheme. Tests undertaken are done within the standards set by the scheme
10. Medical record standards
    Medical records and other reports will be completed in accordance with the district's standards for medical records

F. **Clinical audit**
1. Nursing audit
2. Physiotherapy
3. Occupational therapy
4. Nutrition

G. **Hotel standards**
1. Reception staff training
2. Waiting times for registration
3. Waiting areas:
   Seating
   Catering
4. Bed availability on arrival
5. Cleanliness of wards and public areas — monitoring arrangements.
6. Catering — standards/monitoring arrangements
   — availability out of hours
7. Day room facilities
8. Bereavement room
9. Death registration arrangements

H. **Accreditation**
1. ENB
2. Royal colleges

Source: Offen 1990

Jones (1990) has taken the issues relevant to general surgery further and is a leader in general surgical audit, assisted by the information technology of *Medical data index*. His interest is focused on the fact that in most hospitals, routine information on care such as postoperative complications, unplanned events such as theatre return, readmission for related reason, transfer to intensive care, unexpected deaths, etc, is not readily available. It requires systematic collection facilitated by computer audit systems.

Numerous systems are now available to assist in this process, notably the

Medical Data Index (MDI) utilised throughout the South Western region. Key issues in the developing of information collection systems are:

- Ability to pool information between departments, provider units and regions
- Ability to pool information nationally (eg DANOP Trial of Hospital Infection Control)
- Integration of data collection with daily clinical routines
- Accuracy and completeness of data
- Efficient data capture
- Direct linkage of clinical data storage systems to other hospital systems (eg Patient Activity Database (PAD) and Case Mix Management System (CMMS))

As audit becomes established within the *culture* of a speciality, greater in depth discussion occurs, requiring more sophisticated and differentiated information. Within the surgical speciality several hospitals have investigated the extent to which a small group of diagnoses explain or account for a signficant variance in terms of numbers of patients seen or procedures carried out. Two pieces of research in the South Western region are illustrated in Figures 6.8 and 6.9 and relate firstly to a team of four general surgeons, and secondly to a general surgeon with a special interest in urology.

Procedure	Cases	Occupied bed days
Gastric intubations	622	1115
Cystoscopy	790	1300
All hernias	379	2356
Appendicectomy	390	1695
Mastectomy/biopsy breast	234	1239
Prostatectomy	232	3279
Varicose veins	299	723
Cholecystectomy	254	2097
Sigmoidoscopy	126	904
Oesophagoscopy	96	547
Total top ten	**3420**	**15255**
Total in speciality	5925	37765
% top ten	58%	40%

Comment The procedures reflect a minor aggregation. These results were discussed with the general surgeons. Comparisons with other districts in audit meetings lead to a considerable reduction in the average length of stay.

Figure 6.8 General surgical teamwork load

Procedure	Cases	Occupied bed days
Endoscopic ops on bladder	546	797
Endoscopic ops on outlet of bladder	257	1590
All hernias	97	586
Operations on prepuce	77	37
Endoscopic ops/exams of urethra	76	130
Appendicectomy	66	343
Urethral catheterisation bladder	40	277
Excision and reconstruction of breast	39	128
Diagnostic endoscopic exam of upper gastrointestinal tract	34	242
Cholecystectomy	25	320
Total top ten	**1257**	**4450**
% of all cases	59%	46%

Comment It must be said that this approach is rather simplistic but it does reinforce the point that much surgical work in most provincial DGHS, is concentrated in a few procedures.

Figure 6.9. Urological consultant's workload

Specific audit discussions can focus on particular clinical management decision — taking options such as:

1. Possibility of reducing 25% rate of *normal appendix* appendicectomies. See Figure 6.10.

2. Reducing variability of varicose vein treatment

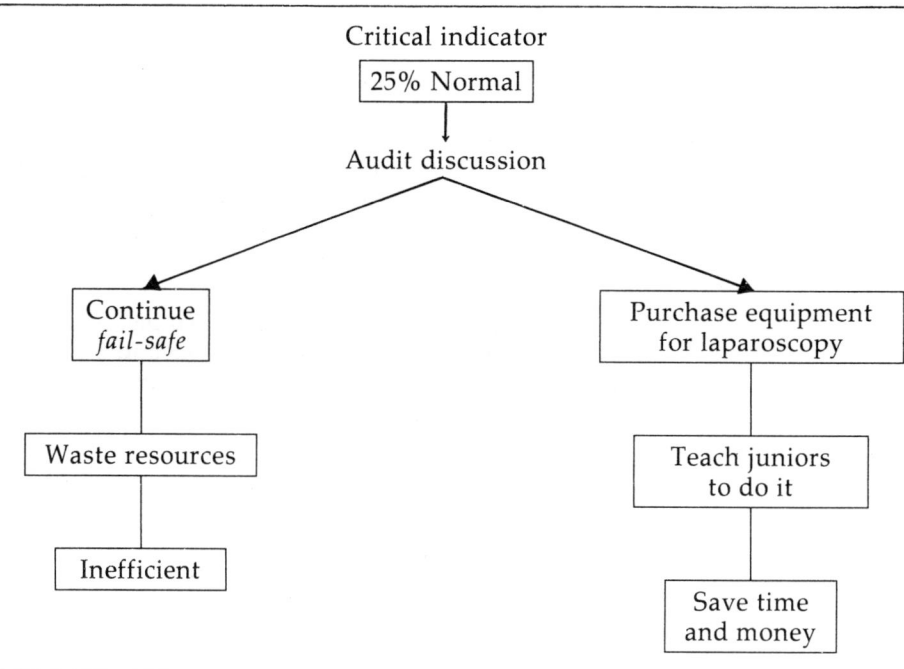

Figure 6.10. Appendicectomy

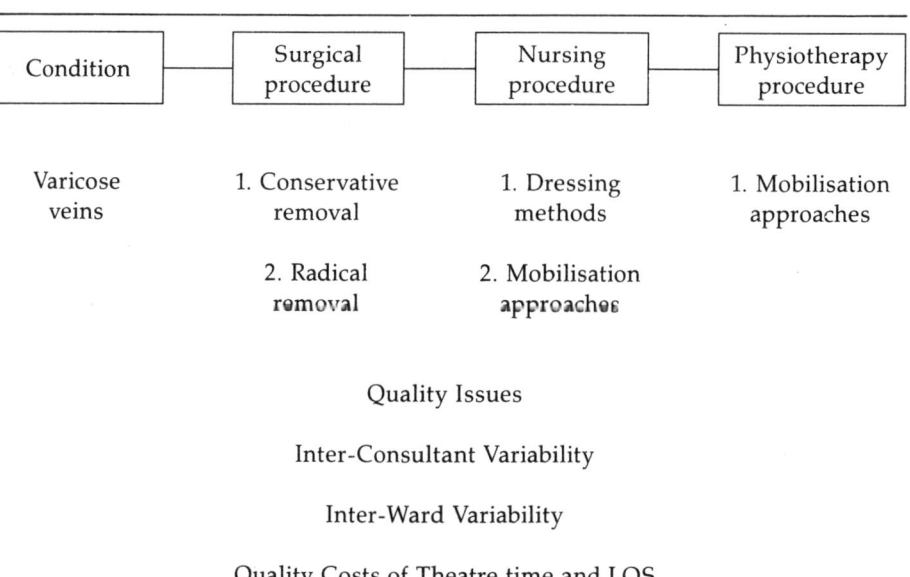

Figure 6.11. General surgery: varicose vein treatment

In conclusion medical audit should be subject to evaluation as it is complex and diverse in the options for investigation and discussion. Doctors, managers, and purchasing agencies should have sufficient information to evaluate the state of audit in a provider unit or particular speciality. A simple and practical audit tool was developed by the South Western Region in its document *Regional Approach to Medical Audit* (SWRHA 1989):

		YES	NO
Structure	There is formal agreement between the governing body, management and consultant medical staff defining responsibility for the review of medical care	☐	☐
	The medical staff corporately accept responsibility for the quality of medical care within the hospital	☐	☐
	There is a consultant or group designated to coordinate medical audit activity within the hospital	☐	☐
	Every doctor working in the hospital is assigned to a formal speciality division of the medical staff	☐	☐
	There is a formally constituted group, including clinical consultants with overall responsibility for:		
	Drugs and therapeutics policy	☐	☐
	Infection control	☐	☐
	Postgraduate medical education	☐	☐
	Medical records (patient information)	☐	☐
	Time for audit is identified in individual consultant programmes	☐	☐
	Timely, accurate local data available to each speciality include:		
	Diagnostic index	☐	☐
	Operations/procedure index	☐	☐
	Mortality listing	☐	☐
	Clerical and computer support is specified for medical audit	☐	☐
Process	Each speciality division meets formally and regularly to review clinical work	☐	☐
	In each speciality, the review:		
	Is attended by all members of the division	☐	☐
	Includes the work of each consultant firm	☐	☐
	Numerically compares patterns of practice	☐	☐
	Recommendations arising from the review are:		
	Recorded for reference	☐	☐
	Distributed to relevant staff	☐	☐
	The general medical staff receives regular audit reports from each speciality division	☐	☐
	The governing board and management receive a regular summary of the process and outcome of audit from the medical staff	☐	☐
Outcome	Recommendations arising from audit lead to documented changes in:		
	Availability of services	☐	☐
	Organisation of services	☐	☐
	Written clinical guidelines	☐	☐
	Clinical practice	☐	☐
	Improvement is measured in agreed key areas including:		
	Patient satisfaction	☐	☐
	Complication rates	☐	☐
	Avoidable mortality rates	☐	☐

Figure 6.12. Regional approach to medical audit (SWRHA, June 1989)

Source: South Western RHA, June 1989

Nursing audit

A major achievement in this area — the NHSME (1990) package *Measuring the Quality* — has already been mentioned. It sets the scene and the 'culture' for nursing care audit in an explicit, understandable, and practical way describing the reasons and benefits of audit; its key principles; and stages of implementation.

Nurses ensure that patients receive a high standard of nursing care. This is *assured* by devising evaluation mechanisms to measure continuously and objectively the *structure, process,* and *outcome* aspects of nursing, against preestablished and agreed criteria indicative of nursing standards (Lewis, 1990). The steps involved are:

1. Identifying the values of expectations of nursing standards.
2. Determining the structure, process and outcome components of nursing practice which are indicative of whether the standards are being met.
3. Developing criteria which are objective and which can be used to determine whether the structure, process and outcome components are acceptable.
4. Developing a tool based on the above criteria to evaluate nursing practice.
5. Developing a method of collecting data to ensure the production of reliable and valid data to be used to determine new practices.
6. Collecting the data.
7. Analysing the data based on nursing policies, procedures and practice.

An overall view of nursing audit is illustrated in Figure 6.13. As in medical audit, nursing audit can be placed in a wide hospital context and can occur at several levels:

- Hospital nursing service
- Ward service
- Individual patient care

Standard-setting methodology has been extensively addressed in Chapter 5.

The three criteria are:

Structure	Physical environmental organisational characteristics which are required. These include equipment, numbers and type of staff, levels of knowledge and expertise, policies and procedures and other resources: what the nurse must have for the standard to be achieved.
Process	Action and behaviour which the nurse is meant to perform: what the nurse must do.
Outcome	Effect of the actions on the patient or client: what the nurse has to be able to demonstrate has happened.

For example for care of patients following abdominal stoma surgery see Figure 6.14.

Individual patient's nursing care

Moving to the audit of nursing care on a more individual patient basis, Individual Patient Audits (IPAs) involve the following:

- Basic information availability on Individual Care Plans
    - General nursing activities
    - Technical nursing activities
    - Administrative activities
    - Time taken for each activity
    - Frequency of activity
    - Skill level of nursing carried out actively.

Clinical audit 75

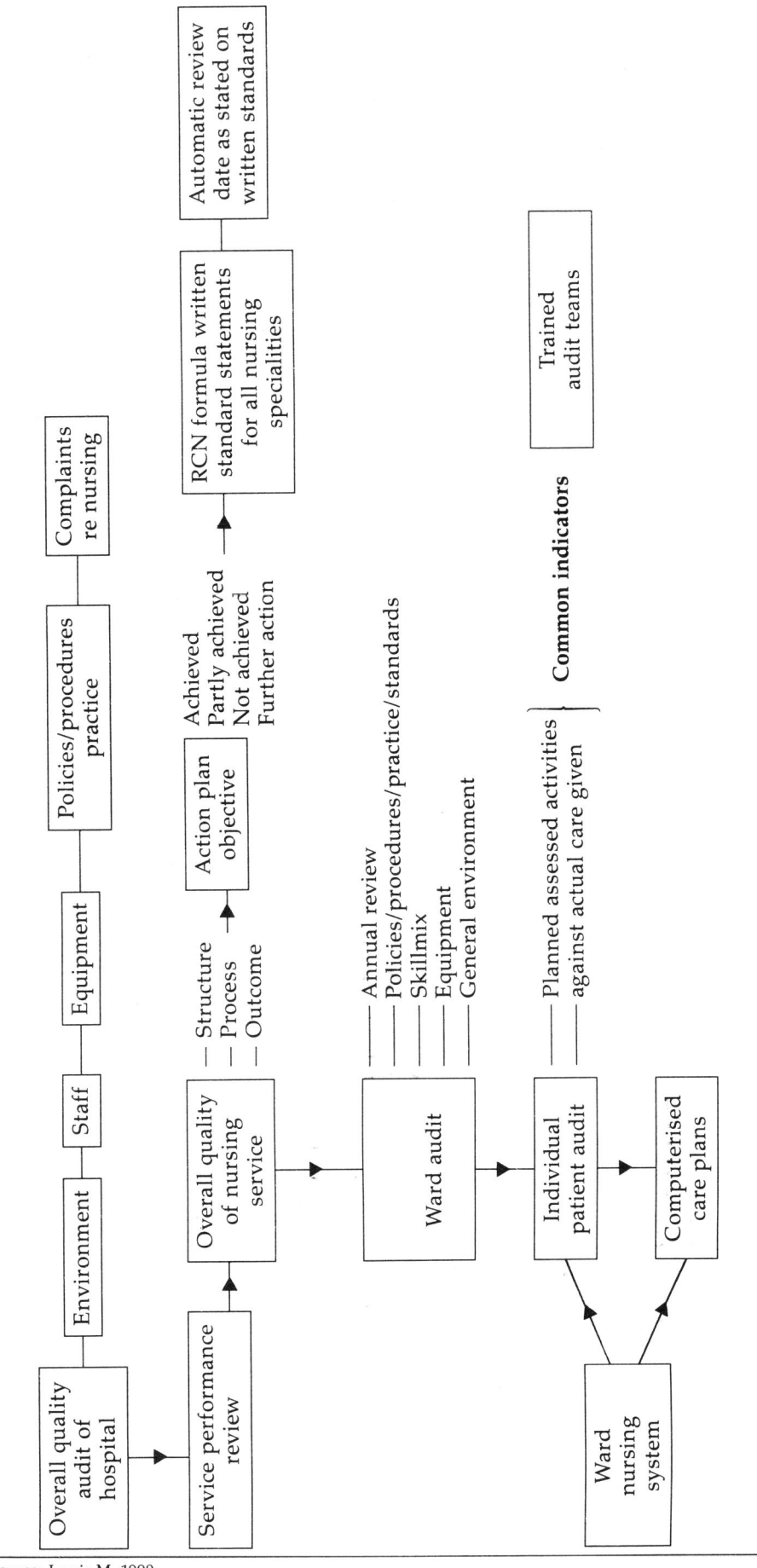

Figure 6.13. Overall view of nursing audit

Source: Lewis M, 1990.

Figure 6.14 Standard care following abdominal stoma surgery

Standard statement: Each patient will achieve self care of their abdominal stoma

Structure	Process	Outcome
1. Suitably qualified nurse available to educate the patient RGN ENB Course Stoma Care	1(a) Patient must have demonstration by nurse of correct technique of appliance management to suit their individual needs	1(a) Patient observes and understands the principles of his own individual stoma care
	(b) Patient participates with nurse in undertaking correct technique of appliance management	(b) Patient confidence and technique improves
	(c) Patient is able to undertake technique with nurse observing	(c) Patient has continued confidence and improved technique
	(d) Patient undertakes total care of stoma unprompted and unaided	(d) Patient confident re self care
2. Privacy	Privacy with guaranteed uninterruption	Patient feels safe and unhurried
3. Appropriate stoma care equipment for type of surgery and patients needs	Nurse should have knowledge of full range and availability of stoma care products	Patient has correct products for his own needs
4. Toilet facilities, washhand basin, scissors, mirror		

Source: Lewis, M, 1990.

Clinical audit 77

Date
Name of patient
Hospital no.   Age   Consultant   Day   Assessed by

Planned interventions

A. Maintain a safe environment
Ensure orientation to ward and ward routine.
Ensure comfortable seating position.
Record TPR..................................hourly
Report if temperature is above 37.4°C.
Observe for pain, swelling or discharge.

B. Communicating
Ensure patient is introduced to personal nurse responsible for today's care.
Encourage patient to ask questions.
Ensure explanations are understood.
Ensure nurse call system available, in working order and within reach.
Indicate if problem with:
sight...
speech..
hearing...
If applicable, ensure glasses are nearby.

C. Pain control
Assess level of discomfort.
Use pain scale where appropriate.
Assist adoption of desired position.
Administer analgesia appropriately and review regularly.
Assess effect of analgesia.

D. Breathing
Encourage coughing and deep breathing.
Position to achieve optimum lung expansion.
Record respiratory rate.
Report if 24 and above, or 15 and below.
Consider health education regarding smoking.

H. Controlling body temperature

I. Mobilising
Assist to enable safe movement from bed to chair.

Figure 6.15. Computerised care plan

Source: Lewis, M, 1990

Ward E   Date 30/3/89   Today's assessment: Assessors initials: HL
Patient number: 2042   MOB   HYG   INCONT   MEALS   PSYCH NEED   CARE GROUP
                          1   1   1   1   1

Basic care assessment	Technical care assessment
Has the patient been assessed correctly at the time of assessment? Y/N [ ]	Is the assessed technical care appropriate? (Y/N) [ ]
(If "NO" please enter correct assessment below)	Have all technical activities been carried out? (Y/N) [ ]
Has the correct basic care been carried out? (Y/N) [ ]	
Mobility (corrected) [ ]	
QA codes [ ] [ ] [ ]	
Hygiene (corrected) [ ]	QA codes
QA codes [ ] [ ] [ ]	Observations [ ] [ ] [ ]
Incontinence (corrected) [ ] (Enter "X" if not incontinent)	Medication [ ] [ ] [ ]
QA codes [ ] [ ] [ ]	Investigation [ ] [ ] [ ]
Meals (corrected) [ ]	
QA codes [ ] [ ] [ ]	Aseptic procedure [ ] [ ] [ ]
Psychological need (corrected) [ ]	
QA codes [ ] [ ] [ ]	Education/counselling [ ] [ ] [ ]
Nursing records	Auditor's comments
Are the nursing records relevant to this patient's care? (Y/N) [ ]	QA codes
	Ward environment:
	Organisation [ ]
QA codes	Staff attitude [ ]
Nursing history [ ] [ ] [ ]	*Patient satisfaction:*
Problem identification [ ] [ ] [ ]	Environment [ ] [ ] [ ]
Goals/objectives [ ] [ ] [ ]	Communication [ ] [ ] [ ]
Evaluation [ ] [ ] [ ]	Support services [ ] [ ] [ ]
	Nursing care [ ] [ ] [ ]
	Auditor's initials

Figure 6.16. Quality assurance assessment form   **Source:** Lewis, M, 1990

- Independent assessor checks values/ratings set against above categories to ensure patient has been appropriately assessed
- Patient and key nurse interviewed
- Comparisons of planned versus actual care
- Feedback to nurse.

An efficient form of individual patient audit (IPA) has been developed in Cheltenham General Hospital since 1986, by Lewis (1990) and colleagues, with the full commitment of ward-based nursing staff. This fully computerised system is linked to the ward (FIP) system and provides full up-to-date computerised care plans (see Figure 6.15) which save nursing time (15 mins. saved per nurse per day); ensure more accurate record-keeping; and provide the *basic information availability* mentioned earlier, which is essential for valid and useful audit.

The Quality Assurance Assessment Form is shown in Figure 6.16.

> This allows an auditor to assess and record for a random sample of patients the quality of care given on a ward. The quality factors are decided upon by each hospital and set in files in the system. The results of the audit can be input back into the system so that an analysis of the results can be made available. The module allows nurses to monitor the service they are providing in order to both consolidate areas of good quality and improve aspects of care which are not reaching the stated standard.

Example of an audit analysis

Quality Assurance/Standards Report

WARD : WARD A

Start Date 6/01/89
Finish Date 6/01/89

Number of patients in sample: 20

BASIC CARE ASSESSMENT

Assessed/administered basic care:
  80% of patients in the sample were assessed correctly
  90% of patients in the sample received the correct basic care

Of the 20 patients assessed:

	Over assessed	Under assessed
Mobility	3	0
Hygiene	0	0
Meals	0	0
Psychological Need	0	0

Incontinence		
Patients assessed as *incont* who were not	:	1
Patients not assessed as being *incont* who were	:	0
Patients assessed as being *incont* but with incorrect assessment	:	0

Basic care QA information

	QA Description	% of sample
Mobility	Patient not mobilised as in care plan	10
	Patient not correctly positioned	5
	Patient not encouraged to self care	0
	Needs for breathing not met	0
	Needs for elimination not met	10
Hygiene	Needs for skin care not met	0
	Needs for mouth care not met	5
	Nail care not undertaken	40
	Patient not offered hand wash before meals/after toilet	15
	General hygiene needs not met	10
Incontinence	Patient not clean and dry	0
	Incontinence not charted	10
	Continence advisor not informed	50
	No evidence of proper assessment	50
Meals	Nutritional intake of patient not recorded	1
	Patient not given the food ordered	50
	Patient not allowed to eat at own pace	5

	Psychological need	Patient not told which nurse is caring for him/her	5
		Nursing staff do not call patient by preferred name	0
		Patient not content with access available to visitors	0

Nursing records QA information

	QA Description	% of sample
Nursing history assessment	No reference to allergies	10
	No emergency contact number	5
	No record of mental emotional state	50
	No record of patient's understanding of condition	75
	No record of relatives understanding of condition	75
	Food likes/dislikes not recorded	10
	Religion not recorded	5
Problem identification	Certain problems of patient not identified	50
	Problem(s) identified medically orientated	10
Goals/objectives	Identified goal(s) unrealistic	0
	Identified goal(s) not patient based	10
	Identified goal(s) not measurable	15
Evaluation	No evaluation of goals evident	50
	Evaluation based on interventions not goals	5

Ward environment QA information

	QA Description	% of sample
Organisation	Ward well organised	100

Patient satisfaction QA information

	QA Description	% of sample
Environment	Not enough privacy been provided	100
	Bathing/toilet facilities inadequate	70
	Ward not quiet enough at night	50
Communication	On admission, not told of ward facility/organisation	0
	Not fully informed of condition	70
	Patient thinks RLAB dissatisfied with communication	100
Support services	Meals have been cold on arrival	70
	Poor quality of food	0
	Food different to what ordered	50
	General dissatisfaction with meal service	50
Nursing care	Nurses kind and polite	100
	Fully involved in care	100
	Nurses responded promptly to requests	60

Other sources of nurse audit material worthy of mention are:

- Monitor — adapted from USA for UK Health Care by Goldstone. (Now adapted for many different nursing specialities)
- Standards of Care and Practice Audit (SCAPA) — developed by Worcester Health Authority in areas of mental health and midwifery (Very practical and useful approach)
- RCN Standards of Care Work — DoH/RCN collaborative project used as the basis for much local standard setting (RCN (1989)) (Currently being evaluated by the DoH).

Clinical therapy services

In 1990 initiatives, complementary to the *Medical Audit Initiative* in *Working Paper 6* were sponsored and funded by the DoH in areas of clinical psychology; occupational therapy; physiotherapy; and speech therapy. This broadening of the professions seen as part of clinical audit was widely welcomed. The first stage of this is to:

- Identify what work has already been done on audit within these professions
- Consider the most appropriate next steps, especially in relation to the feasibility of multidisciplinary approaches to clinical audit involving these professions.

The second stage is to:

- Develop suitable audit tools on individual profession or, hopefully, multidisciplinary basis
- Realise benefits for patients from audit in this/these areas.

The output from this exploratory research is awaited. Meanwhile, the many aspects of clinical audit described in this chapter have varying applicability to the clinical therapy services. But standard-setting is beginning to gather momentum in most provider units.

Clinical psychology has been addressing the issues of quality assurance and clinical audit since mid 1989 when its draft guidelines *Quality Issues in Clinical Psychology* were produced (BPS, 1989). It was felt that four levels of *interface* were worthy of exploration by psychologists in evaluating and auditing their work:

- Client/Practitioner Interface
- Practitioner/Multi Disciplinary Speciality Interface
- Speciality/Provider Unit Interface
- Provider Unit/Purchasing Agency Interface.

The Donabedian categories of *structure — process — outcome* were complemented by a category labelled *congruence*, addressing the *harmony* existing between a client and his/her psychologist and also a category *quality assurance* reflecting time and support for audit and quality assurance. These interfaces and categories are frameworked in the matrix shown in Figure 6.17.

Dimension/ Interface level	Congruence	Structure	Process	Outcome	Quality assurance
Client/ independent practitioner Interface					
Independent practitioner/ service speciality Interface					
Service speciality/ district Interface					
Interface with external/ higher level forces					

Figure 6.17. Clinical psychology framework

Source: BPS, 1989.

Each cell in this matrix is used to provide a comprehensive basis for clinical psychologists to measure performance against defined standards.

Here is an example of issues being addressed at client/practitioner interface mentioned in Figure 6.18.

Congruence	Structure	Process	Outcome	QA processes
Practitioner share values	Setting accessible and blends with locality	Access to the service/ psychologist	Agreed goals achieved	Clear locally agreed and relevant standards
Choice of treatments	Setting offers privacy and confidentiality	Appropriate assessments	Client satisfaction with: settings; outcome; style of approach	Routine information gathering for monitoring
Shared/agreed theoretical base for treatment	Setting enhances respect and dignity	Validated and effective interventions	Impact on quality of life	Self/peer review of processes
Shared formulation	Setting takes account of special needs	High quality interpersonal skills	Care is coordinated	Monitor progress of treatment
Shared objectives	Necessary equipment is available	Flexibility of approach		Records organised for analysis
		Regular reviews of progress		Client satisfaction surveys
		Relationships/ sharing with other staff coordination		Formal allocation of time for QA
		Clear records for all elements of treatment/ intervention		Could outcomes have been achieved more efficiently or effectively?
		Shared decision making with clients: formulation and objectives		
		Acknowledgement of referrals and initial client contact/ appointments		
		Work reflects theory		

Figure 6.18. Clinical psychology QA working party – Dimensions of quality at client/practitioner interface

Using a *contracting framework*, work has recently been completed at the Health Services Management Centre in Birmingham, on specification formats for *Professions Allied to Medicine* (HSMC, Series 18, 1989). In this document, key outcomes and standards of performance for occupational therapy and chiropody were identified:

Occupational therapy

Key result areas	Standards of performance
1. Patient functions Achieving/maintaining the highest level of personal functioning possible	X number of patients (with A condition?) will achieve/ maintain highest level of function within Y time

2. Prevent readmissions
Minimise the need for re-admissions to secondary care by ensuring initial discharge is safe and appropriate

No more than X% of patients for whom OT provides services are readmitted to hospital because of inappropriate discharge plans

3. Facilitate discharge from hospitals
Help minimise length of patient stay in hospitals

X% of patients referred to OT discharged from hospital within Y weeks, in a period of six months

Chiropody

Key result areas
1. Access
Maximum access of population who benefit from chiropody services

Standards of performance

95% of new patients should be seen for assessment within four weeks and receive first treatment within two weeks of assessment

2. Mobility, elderly population
Maintain mobility of elderly population within the community

X% of over-65 population should be seen by the service in any one financial year (currently 23%)

3. Preventive, foot problems
Prevent occurrence of unnecessary foot problems within the community

a. Ratio of provision of orthoses (appliances) to the desired prescription rate of orthoses
b. 75% of 10 year olds should be screened at school for potential foot problems

Integrated clinical audit

Considerable achievements are now being established in each of the clinical professions in terms of establishing key components of professional audit:

- An *Audit* culture (aims and core values)
- Simple methodology — how to start
- Sophisticated/complex methodology

The rates and consistency of implementing these components, like all NHS initiatives, is variable between professional groups and between provider units. However, many clinicians face a dilemma when planning audit, which is illustrated when he/she considers:

- The patient-view of services
- The clinician's view of how his/her service **actually** functions.

The dilemma is brought about by the patient's path or trail through a service cutting across professional boundaries. In a good quality service this is relatively easy and efficient: the nurse or doctor providing care liaises closely with other clinical therapy colleagues providing *seamless* service. Key aspects of the typical *patient trail* followed in a general hospital is shown in Figure 6.19.

An approach to clinical audit to harness and integrate the efforts of multidisciplinary clinical, and support, teams would be to discuss each of the eight stages along the patient trail. Relevant clinical **team** members would be invited to discuss the many quality issues, both clinical and social, which relate to the appropriate stage of patient care and service. The seven TQM features are the obvious areas to initially investigate.

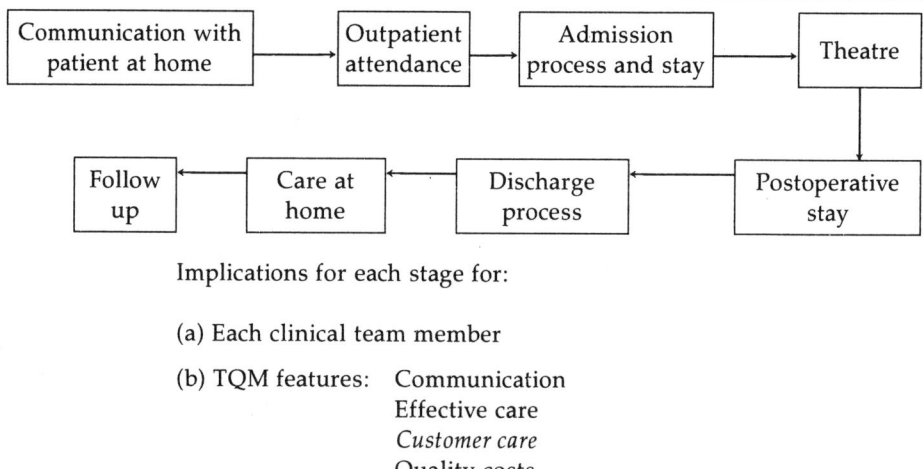

Figure 6.19. Patient trail

In **clinical audit** terms, this *integrated* approach is gaining considerable support both at hospital and professional level and also, as mentioned earlier, at DoH level. These groups realise that individual patient care and service are usually provided by a corporate team and audit should reflect this delivery process. An example of a *patient trail* for someone with minor injuries is shown in Figure 6.20.

Integrated clinical audit means:

- Involving medical, nursing and PAM professions in audit strategy
- Linking audit focus to *patient trail* and *patient* health care outcomes.

Another *patient trail* involving elderly service rehabilitation illustrates these points as in Figure 6.21. A practical and successful example of this was initiated by Brighton Health Authority in conjunction with CASPE Research Group using *Occurrence screening* in hospitalwide clinical audit. It was a method of multidisciplinary, hospitalwide clinical audit involving:

- Systematic identification and analysis of events or *occurrences* during a patient's treatment which indicated less than appropriate quality of care/service
- Analysis of causes and effects of each occurrence
- Information from analysis used to plan changes in clinical practice.

This method has been extensively used in Australia, Canada, and USA, where it was first developed in the mid-1970s.

Screening criteria are designed to identify events which may indicate poor quality. They are selected by any member of the clinical team either from information obtained externally or from their own experience.

Patients' medical and nursing records are screened on discharge against set criteria for *occurrences* matching one or more criteria. If found, these records are passed to the audit team for discussion on effects, avoidability, and possible changes in clinical practice. The database which can be produced using this methodology allows the interrogation of the occurrence screening system to look at areas such as:

- Listing of all cases of unplanned return to theatre over last six months and avoidable reasons for return

Clinical audit 85

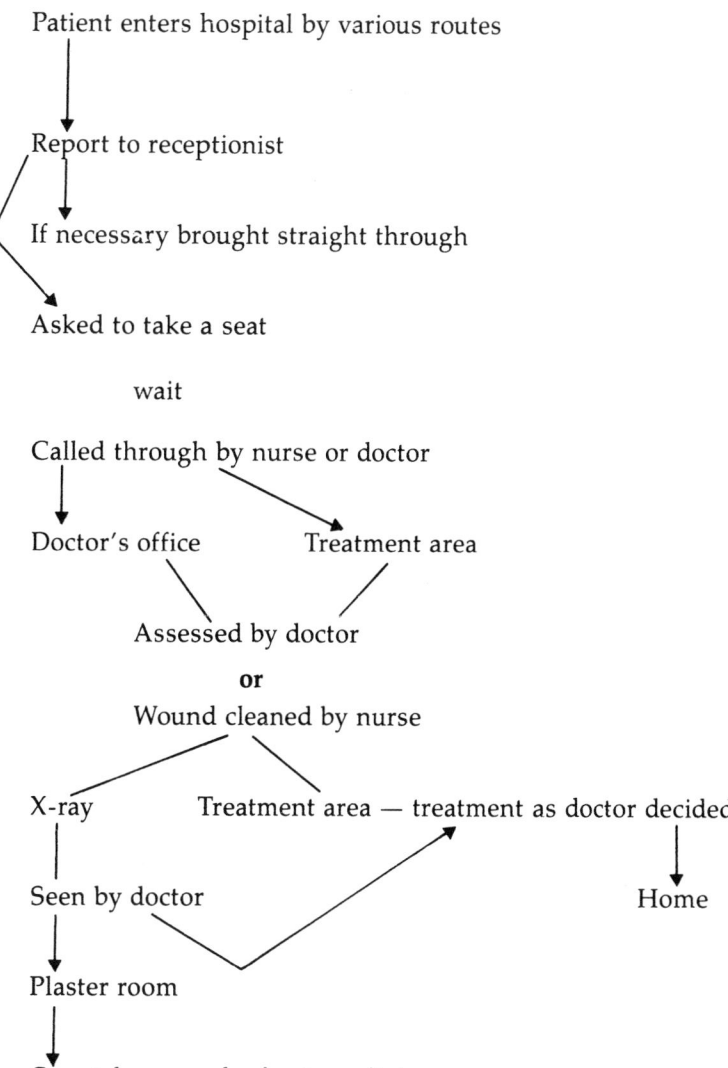

Patients with minor injuries

Patient enters hospital by various routes
↓
Report to receptionist
↓
If necessary brought straight through
↓
Asked to take a seat

wait

Called through by nurse or doctor
↓
Doctor's office → Treatment area
↘ ↙
Assessed by doctor
or
Wound cleaned by nurse
↙ ↘
X-ray   Treatment area — treatment as doctor decided
↓ ↓
Seen by doctor   Home
↓
Plaster room
↓
Care taken over by fracture clinic

Figure 6.20. Specific patient trail

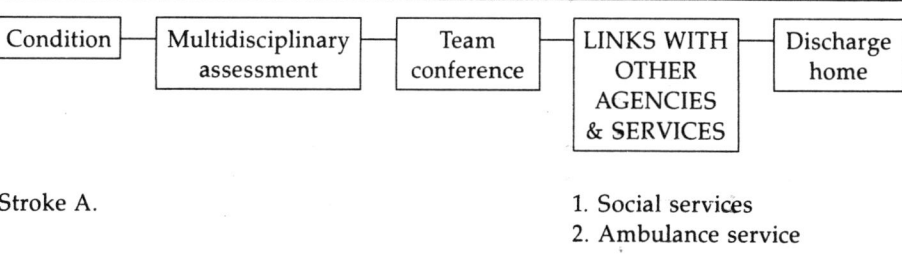

Stroke A.

1. Social services
2. Ambulance service

Quality issues

Need for multidisciplinary assessment
Integration of geriatric and general medical care
Drug expenditure
AIDS provision
Day hospital/care alternatives
Intensive care and non-intervention
Readmission rates
Length of stay and relative outcome

Figure 6.21. Elderly service: Rehabilitation of stroke patient

- Listing of patients undergoing abdominal surgery who subsequently develop wound infection. Does this incidence vary between surgeons and/or wards?

Key evaluative questions are whether this method of audit provides information for clinicians in a useful form and if the information can be translated into changes in clinical practice where appropriate.

A summary of generic screening criteria and known exceptions is given in Figure 6.22.

Criteria title/description	Notes on known exceptions and on interpretation
Admission for adverse results or complications of outpatient management	Except planned/expected admissions
Readmission for complications or incomplete management of problems on a previous admission	Except planned/expected readmissions
Error in obtaining consent to operative procedure	Except emergency surgery, where patient cannot give consent, or life threatening problems found and addressed during surgery
Unplanned removal, injury or repair of organ or structure during surgery/invasive procedure	No exceptions
Unplanned return to theatre	Except returns for planned second procedures
Pathology/histology report varies significantly from preoperative/antemortem diagnosis	No exceptions. Check for variances which have resulted in a significant difference in the patient's care or treatment
Transfusion problems: reactions, complications, improper usage	No exceptions. Check for problems not addressed appropriately
Hospital acquired infection	No exceptions. Document type/nature/effect of infection
Antibiotic/drug utilisation problems	Check for medication errors, anaphylaxis, and deviation from mandatory therapies/regimes
Cardiac or respiratory arrest	No exceptions. Document handling of arrest, and circumstances leading to arrest
CVA or acute MI within 48 hrs of surgical procedure, or PE at any time postoperatively	No exceptions. Document circumstances leading to incident
Unexpected transfer to higher dependency unit	No exceptions. Document cause/circumstances of transfer
Specified patient related clinical complication(s) occurred	Check list of specified complications: urinary retention, IV problems, inadequate pain control, pressure area development, DVT, and so on
Specified patient related non-clinical problems	Check list of specified problems: theatre booking cancelled or delayed, no ITU bed available, patient placed on clinically inappropriate ward, casenotes/records missing, patient had slip or fall, and so on
Neurological deficit on discharge not present on admission	No exceptions. Check particularly for neurological deficits resulting from surgery
Unexpected death	Except patients receiving terminal care/category C patients
Medical record review	Check for specified deficiencies in medical records
Nursing record review	Check for specified deficiencies in medical records
Evidence of patient/family dissatisfaction	No exceptions. Document cause and handling of dissatisfaction
Discharge related problems	Check list of specified problems: discharge delayed for non-clinical reasons, no written evidence of discharge planning, and so on

Figure 6.22. RSCH/SEH Occurrence screening project: Summary of generic screening criteria

Source: Criteria developed by: CASPE Research/Brighton Health Authority Department of Public Health, Brighton General Hospital, Elm Grove, Brighton, BN2 3EW; Tel (0273) 696011 Ext 3096

Note: Abbreviations: Cerebrovascular accident (CVA); Deep vein thrombosis (DVT); Intensive treatment unit (ITU); Intravenous (IV); Myocardial infarction (MI); Pulmonary embolism (PE)

Occurrence screening is well-suited to integrated clinical audit as the criteria can apply to medical, nursing, and paramedical care leading to appropriate review mechanisms for either individual professions or as a group. The latter reduces duplication of effort inherent in a series of audit systems relating to individual professions (Bennett and Walshe, 1990). It will lend to identifying quality issues which cross professional boundaries. Integrated clinical audit reflects the increasing collaboration between clinical professions, notably medical and nursing staff. It endorses the development of *services* with defined, explicit outcomes, rather than the continuation of professional/ departmental development and growth as being the indicator of *success* in provision of care. Collaborative Care Planning (CCP) which is being developed in many US hospitals, combines nursing care plans with medical condition — specific protocols. The single set of care plans for an individual patient are monitored and updated as treatment progresses. As a whole the care plans are regularly reviewed and audited to highlight issues of effectiveness and efficiency, covering all aspects of medical, nursing, and paramedical care. This model has implicitly operated in many hospitals and community services and explicitly in some. It certainly has potential for greater use in the NHS.

A variation of the theme of collaborative care planning has always existed to a greater or lesser extent in the mental health field, encompassing as it does, NHS and Social Services provision. As a result of *caring for people* legislation, a form of CCP is being implemented — namely, *case management.* It focuses on the mix of services and the way these services relate to each other rather than on the individual service. Systematic practice of case management involves high quality components of:

- Engagement
- Assessing
- Planning
- Accessing resources
- Coordinating
- Disengaging.

Although it is said that case management is about the administration of services and does no more than facilitate access to services, it is in itself a way of organising a quality approach to cross-agency collaborative care and, as such, is an important area to audit in terms of its effectiveness and efficiency. It should be evaluated in terms of its success in:

- Identifying appropriate cases
- Allocation of key worker
- Management of resources

and tangentially
- Providing effective care.

Conclusion

Clinical audit, whether it be on an individual professional basis or more integrated/multidisciplinary must realise the benefits mentioned in the opening section. It has to achieve:

- Greater consistency of care and overall quality — Provided both by individuals and multidisciplinary teams
- Reduction of clinical or team coordination errors — On the basis of the best available knowledge and insight, audit must reduce the *risk* of care and avoidable problems

- Improved cost-efficiency — Any opportunity, via audit discussions to reduce cost, making savings for reallocation to desired developments is essential
- Improved clinician confidence — All clinical staff need audit to help increase their awareness of their skills and limitations at personal, professional, and corporate levels.

Chapter 7 Communications: getting the message across to patients, staff, and teams

The range and complexity of quality improvement activity in provider units means that, by necessity, all staff must be involved. At each level of the unit, many of the strategies and associated issues are best managed through teamwork. Effective communication processes, whether applied to individuals or teams, are essential to total quality management. Quality communication ensures that problems are: tackled by large numbers of staff, maximising the unit's capabilities; investigated by harnessing the range of knowledge, skill and experience of many staff; addressed across professional boundaries and boost *team* morale.

Effective communication involves:

- Establishment of an effective strategy for ensuring good communications
- Establishment of effective internal procedures eg team briefing, communication, audit, action teams
- The establishment of effective external procedures eg patient information, customer feedback
- Gaining commitment at all levels in a unit for the unit's plans
- Development of relevant communication skills.

In April 1990, the chief executive of the NHS Management Executive (NHSME) launched the new NHS Communication programme aimed at improving the awareness and understanding among NHS staff of major changes within the NHS and how they would affect jobs and patient services.

Figure 7.1. Management executive objective

Help raise the overall standard of management by measurably improving communications, both within units and between units and the rest of the NHS

Source: Department of Health

During several meetings in all fourteen NHS regions, the Chief Executive Duncan Nichol explained the importance of communication helping to give the service a sense of direction. All managers and staff have to share and understand the specific priorities for action in the NHS:

- Good financial control
- Maintenance of service levels through manpower efficiency
- Dialogue with GPs to identify referral patterns
- Analysis of strengths and weaknesses of existing services
- Basic information systems to support contracting.

Effective and well executed communication strategy, will facilitate the considerable NHS agenda.

The initial part of the communications programme was a survey of all Unit General Managers in England as to their views of current existing communication processes between DoH, Region, District, Unit, and Sub-unit levels. The findings of the survey questionnaire were discussed with DGMs and UGMs in regional meetings.

Survey findings showed that nationwide 99% of UGMs felt committed and responsible for helping to build the unit's reputation, and 87% believed in the changes they were implementing. However, many felt there was a perceived lack of trust which hindered managing the changes.

This problem lay between managers and staff. Approximately 50% of UGMs felt they did not always receive accurate information *from* their staff, and that information received from *above* became increasingly difficult to disseminate to staff, the greater the distance from the unit (from district 38%, from region 17% and from DoH 13%). Timeliness of information receipt was also a problem.

It was also found that many UGMs felt their managers, especially middle managers, needed more training in communication issues. In practical terms most managers perceived a difference between preferred communication method (informal and personal) and communication practice (structured and written).

A summary of key messages from the survey is:

- UGMs are extraordinarily committed to the job
- But there is a problem about trust
- Managing change demands better communications
- Improvements are in hand, but need validating — especially with staff
- The fundamental need is to raise the confidence and skills of managers in the unit.

Source: Department of Health (1990)

As a result of this survey, the NHSME identified the following requirements in order that communication improved in provider units:

> Attitude survey and communication audit of staff in provider units
> Corresponding programmes for commissioning agency staff
> Workshops on communications for UGMs to provide the following outcomes:
> > Build a greater understanding, commitment and skills of UGMs and their teams
>
> Establish staff attitudes and patient perceptions
> Prepare and implement unit communications plans and measure outcomes — the level of improved communications.

It is a salutary lesson to note that even the *best* run into communication problems themselves in that the *best laid (NHSME) plans* went slightly awry at this stage.

The first set of management workshops had variable success due, in part, to the difficulty of clearly understanding the level of communication skills already evident among the UGMs in order that the training/continuing education provided was appropriate to their levels of competence. As a result of this, the NHSME plan was modified and regions themselves were given greater flexibility to achieve the outcomes listed above via their own methods — an approach with some advantages.

Provider unit communication improvement

The main steps towards improving communications within a provider unit are illustrated in Figure 7.2.

```
                    ┌─────────────────┐
                    │  Communications │
                    │     strategy    │
                    └─────────────────┘
            ↙        ┌─────────────────┐        ↘
┌───────┐            │  Communications │            ┌──────────┐
│ Staff │            │      audit      │            │ Patients │
└───────┘            └─────────────────┘            └──────────┘
    ↓                                                    ↓
┌──────────┐         ┌─────────────────┐            ┌──────────┐
│ Internal │         │  Communication  │            │ External │
└──────────┘         │   improvement   │            └──────────┘
                     │ implementation  │
                     │      plan       │
                     └─────────────────┘
                          ┌──────────┐
                          │ Actions/ │
                          │ Results  │
                          └──────────┘
    ↓                                                    ↓
Communication techniques                 Quality information to patients
Communication skills training            Customer feedback
Team work enhancement
Standard-setting, monitoring review
```

Figure 7.2. Improving provider unit communications

Provider unit communication strategy

As with all deeply held strategies, the greatest success is obtained with a communications plan when the majority, if not all, staff are involved with constructing and contributing to the *strategy* that precedes the *plan of action*.

At a recent communications strategy workshop with NHS managers, the author facilitated the constructing of the strategy shown below. Its core values are self-explanatory and illustrate the range of issues which need to be addressed when considering communications improvement.

Communication strategy

Mission statement

Communication processes need to be commonly understood by all staff and meaningful to both patients and staff. Communication should be efficient, effective, honest, accurate, timely, and relatively undistorted. It should reflect a respect for the value for staff and reinforce the corporate objective of the highest quality of health care delivery. Standards for good communication should be measured, monitored, and improved upon.

Core values

1. The quality of the message should be high in terms of clarity, relevance, timeliness, etc and reinforce the main objectives of the service ie to provide high quality care and developments in care. Time and accessibility must be addressed.
2. There must be a facility for two-way communication between all *consumers* and *suppliers* whether between staff and patients, or between staff and staff
3. Communication should demonstrate a value for staff and their skills
4. Effective customer communication at the *shop window* should be given a high priority in any health care service
5. There will be a commitment given to training and staff updating in communication skills
6. A system for establishing measurable standards in communication, processes for regular monitoring, and review of standards and conformance to these standards.

Communications audit: staff

Prior to establishing processes for **improving staff communication** within a unit, it is important to:

> Understand the staff's view of the climate in which communications take place, in order to provide managers with information directly relevant to the management of change
> Obtain staff's view on specific communications issues
> Use the data to give indications about which areas should be given priority attention
> Provide benchmark data upon which a communications plan can be based, and progress subsequently measured.

Sampling can be random, representative or total, and is usually confidential. Key areas covered in such a survey/audit include the following areas (FPC Enfield/Haringey, 1990):

1. **Goal alignment:** Everybody in the organisation is pulling in the same direction — individual, unit and department goals all relate to the organisation's mission and strategic vision
   Benefits: Builds cohesion, efficiency and effectiveness of all activities; builds confidence in management; provides direction for planning and individual efforts.
2. **Good communication:** Clear, sufficient and appropriate information flows top-down, bottom-up, and laterally throughout the organisation. All communications are read or watched and understood.
   Benefits: This condition is a necessary requirement for establishing an appropriate organisation culture. It is also a common theme running through all areas.
3. **Initiative:** Employees have the opportunity to exhibit the willingness to think for themselves and act independently in helping an organisation address problems in areas where they can reasonably make a difference
   Benefits: Provides a competitive edge for innovation and productivity, while discouraging inertia; increases the unit's capacity to adapt and respond under pressure; provides opportunities for self esteem.
4. **Team work and trust:** People talk and work together in functional and multi functional groups with cooperation and honesty
   Benefits: Improves organisational efficiency and effectiveness; promotes goal alignment; builds trust and respect amongst colleagues; reduces politicking and excessive pigeon-holing.
5. **Adult treatment:** Each employee, regardless of their level or function, is treated with dignity and respect. Controls and rules make sense and are not seen as unnecessarily demanding
   Benefits: Prevents feelings of anger and frustration that can create withdrawal of commitment. Encourages teamwork, initiative, pride, and confidence.
6. **Personal development:** Employees have sufficient opportunity to develop new skills and gain exposure by new responsibilities and tasks. Employees are encouraged to develop themselves, as people and professionals, within and outside the day-to-day organisational environment
   Benefits: Meets the particular employment needs of this generation; prepares people to be more effective; builds optimism about the future and therefore enhances commitment.
7. **Service/reputation:** All employees possess a sense of kinship with patients and a deeply felt understanding of their needs; they understand the organisation's services and their work is perceived as ultimately driven by patient need
   Benefits: Creates staff who are more likely to feel loyal and committed.
8. **Natural work ethic/pride:** Employees are happy and comfortable with their jobs and in their job environment; they have pride in the

organisation and in their work

Benefits: People who are comfortable in their environment are more effective, efficient and productive; natural work ethic promotes innovation, pride, and a progressive spirit.

From inspection of survey results, key areas for further investigation will be evident and require senior managers to clarify and make recommendations for solving particular communication problems. Taking area item 2. *Good communications* in greater detail, key issues for investigation would be:

A. How are staff most likely to receive information?
   1. Written
   2. Newsletter
   3. Notice board
   4. Uni-disciplinary meetings
   5. Multidisciplinary meetings
   6. Union
   7. Newspapers/TV
   8. Grapevine
   9. Professional journals
   10. Team briefing
   11. One to one
B. To what extent does the information cover?
   1. The unit's future plans
   2. Performance against budget
   3. Quality and quantity of clinical service provided
   4. Patient satisfaction
   5. Investment in new wards and equipment
   6. Changes in working practices
   7. New legislation
   8. Pay conditions and employment
   9. Likely changes in job description/roles
C. What communication style is evident at senior management level?
   1. Informal – formal
   2. In person – written
   3. Autocratic – democratic
   4. Open – hierarchical
   5. Dyadic – groups
   6. Standard techniques – idiosyncratic
   7. Involving unions – avoids union involvement
   8. Open-door – *appointment only* policy

Communications skills training

Working relationships within the NHS are based on effective communication skills.

Communicating with patients

Health service professionals whether general managers, administrators, nurses, doctors, professionals allied to medicine, or receptionists, encounter many situations in which a high level of interpersonal competence and confidence is required. Patients must be dealt with in a sensitive and caring manner, especially if they are anxious, upset, or even mildly irritable. Difficult issues can arise which require delicate handling.

Communicating with colleagues

In providing essential health services to the public, staff of all professions need to be able to communicate well with each other. Positive management is based on good communication involving listening, recognition of feelings, and positive influencing. These skills are necessary to many different aspects of management and interpersonal relationships. The following areas are of particular relevance:

1. Managing subordinates: This often requires a high level of flexibility from managers and involves being able to listen carefully, recognise employees' feelings, and question them appropriately to get relevant information. It necessitates being able to influence in a direct and specific manner.
2. Appraisal: When this occurs on an annual or more regular basis, it is essential to be able to give feedback which is both specific and constructive, whether it be positive or negative. It should lead to concrete changes in the behaviour of the employee.
3. Warnings and disciplinary action: When situations get difficult and objectives are not met, it becomes necessary to give warnings concerning future behaviour. No one enjoys this. However, there are ways of doing this which are both effective and lead to the greatest chance of difficult situations improving.
4. Introduction of new staff: When new staff join a unit, it is a golden opportunity to *model* good open communication and instil the confidence that they will be listened to and treated in a friendly and effective manner.
5. Counselling subordinates with emotional problems: When one of our colleagues has a problem which affects his/her work performance, managers need, at times, to be able to offer help, in terms of listening and being sympathetic and, at times, offer effective guidance. We, ourselves, may need this counselling from our own managers.
6. Being influential in meetings and small groups: Most workers spend some of their time attending meetings or being members of small working groups. When group members are assertive and try to be cooperative and influential, meetings can be highly successful and rewarding. To do this, one has also to be able to confront difficult situations or ambiguous mixed messages.
7. Team building: Increasingly, multi-professional teamwork is part of the NHS. This involves a high level of interpersonal competence and flexibility, knowing when to: listen; give advice or information; be assertive; be confronting.
8. Shop window reception: Those such as receptionists and secretaries who meet clients or patients in out-patient clinics or health centre settings need good *shop window* skills ie being polite, cheerful, friendly, as well as being helpful. To do this several hours a day non-stop is a highly demanding, albeit necessary, job.

Who could use better communication skills?

Unit general managers and unit officers will regularly meet senior staff within their unit. If they show they are willing to listen, use time effectively, be influential and useful when they do speak, this will encourage other unit staff to communicate clearly and effectively.

Consultants, senior nurses and heads of departments are often in contact with a large number of any unit's workforce on a daily basis. Whether this contact be with clinical or support services, a high level of interpersonal competence is required.

Ward nursing staff run busy wards and meet many patients and their relatives, often under stressful and frustrating conditions. To be able to rely on good communication skills means effective care is more likely to occur.

Receptionists and secretarial staff need good *shop window* skills. Most consumers of the Health Service often mention how they were *treated* at the clinic door as well as their overall care.

Learning better communication skills

Some people feel that communication skills are something which one has or hasn't got, something which is not learnt. Some feel it is secondary to forceful management. In reality, everyone needs to improve their abilities to communicate clearly with others, at whatever level of the Health Service system they work. As a manager, one cannot be *forceful* successfully without being able to listen, encourage, be understanding, and able to influence others in several different ways. These skills are summarised in Ivey's *Microskills Hierarchy* illustrated in Figure 7.3.

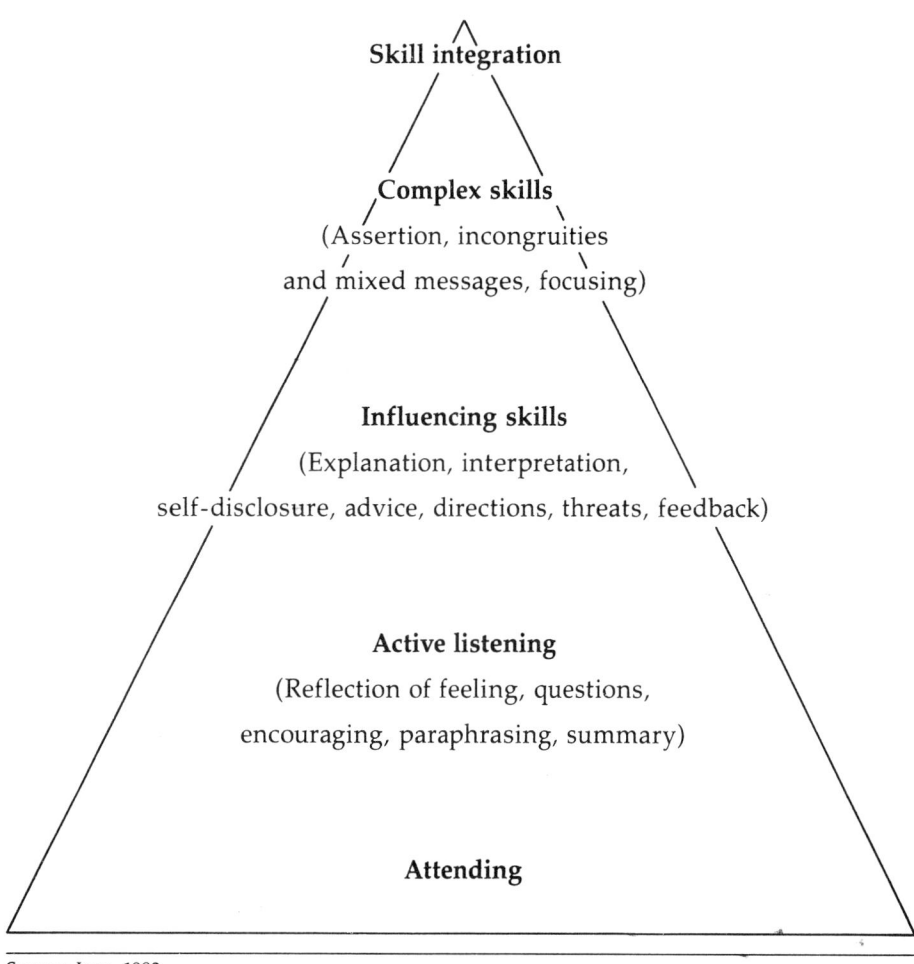

Figure 7.3. Microskills hierarchy   Source: Ivey, 1983.

Workshop based training (eg Face to Face training, Chartwell-Bratt) can help to achieve a greater level of communication skills and also ability to train others in these skills. Effective workshops utilise:

Information giving
Observations of good and bad models
Videos
Role playing rehearsal of individual/micro skills
Constructive positive feedback

96 Total quality management in health care

	Unit senior managers	Medical staff	Middle managers	Ward level nursing staff	PAM staff	Non clinical staff	Unions	
	\multicolumn{7}{c}{Levels of communication}	
Complexity, geography and size								Raise priority of communication
Local history of district/unit	P						S	Meetings with staff
	R						O	
	O						L	Newsletters
Recent management changes	B						U	
	L						T	Make information relevant etc
	E						I	
	M						O	
Relations with RHA	S						N	
							S	Increase interpersonal effectiveness
Lack of relevant and timely information								
Management style								Encourage ownership of communication
Lack of consultation								Develop training
Poor recognition of issue at patient care level								Positive management style
Communication with night/part time/agency staff								Audit and review current communication patterns
Accessibility of senior staff								*Ad hoc* techniques
Interpersonal effectiveness								

Figure 7.4. Levels of communication

Source: Hugh Koch

Communications improvement implementation plan: internal

As a result of the internal unit communications audit, particular problems will have been identified, operating at certain levels/certain professional groups in the unit, with a potential number of solutions. These are illustrated in Figure 7.4.

Four particular methods or techniques of improving communications within a provider unit are discussed below as arguably they have the greatest effects on improving effectiveness of communications in their own place. They are:

> Team briefing
> Open forum meetings
> Unit newsletter and
> TQM information.

Team briefing

This approach to ensuring key information is disseminated to **all** staff within an organisation is well tried and tested in many large organisations although it is not without its pitfalls. The model most widely used is adapted from the Industrial Society approach and is defined as:

> a formalised system for passing messages from management to all parts of an organisation through a structured but flexible approach.

It involves team leaders, usually line managers, meeting with their staff on a regular basis. The basic information is passed from top management: either district or unit senior management, the latter becoming increasingly more appropriate as the starting point, ensuring **all** staff, including night staff and part-time staff, are briefed within a certain period of time (eg 72 hours, one week, four weeks).

Information communicated is in two, or more parts as in Figure 7.5.

1.	Core brief	originating from senior management eg chief executive, unit general manager
2.	Relevant brief	decided by next-in-line managers eg hospital manager, assistant general manager, director of operations.

The relevant brief can be further repeated by the next level of management/supervisory staff if they exist before finally reaching the direct/care service staff. An effective innovation to the team briefing process is the final *column* or part of the process whereby staff are encouraged to feedback any comments appertaining to the Core and/or relevant brief(s) to the originator, the chief executive/unit general manager. An example from one acute unit is given in Figure 7.6.

This approach is very advantageous provided the following issues are taken into account:

> Team briefing requires preparation and training to ensure key managers understand the process
> Team briefers, the line managers, must be committed to the strategy.
> Lip service is quickly noticed and is catastrophic!
> The team briefing process should be regularly monitored and reviewed.

What are the positive outcomes? Team briefing improves communications (Silverstone 1990). The preparation of staff who feel they are kept fully or fairly well informed by unit management increases (25-50%) and the

Month:

Core brief	Relevant brief	Hospital	Head of department/ middle manager notes	Staff feedback to unit general manager

Figure 7.5. Acute unit: team briefing

Communications 99

Core items	Details	Hospital	Head of department/middle manager notes	Staff feedback to unit general manager
1. Project 2000	We have started to recruit health care assistant trainees. If anyone knows of someone who might be interested, please contact..........the RGN course has been discontinued. Multi-disciplinary issues are being set up to keep staff updated			
2. Quality management	Attendance at the open forum meetings has been good with the following hospital attendance figures: Hosp A – 80, Hosp B – 80, Hosp C – 50, Hosp D – 18, Hosp E – 15. Would those who attended please let colleagues know what was discussed. I and unit officers would be very interested to hear what ideas came from your own meetings. The best idea I receive for your quality improvement in February will earn a £25 prize for its *owner* — let me know!			
3. Space utilisation meeting (Hosp A)	Planning has now been almost completed – we're still waiting for news on funding! Many thanks to all involved.			
4. Open forum meetings	Friday 9th Feb 2.15–3.15 Hosp A (Blue Dining Room) Tuesday 13th Feb 1.45–2.45 Hosp B Wednesday 14th Feb 10.00–11.00 Hosp C Wednesday 14th Feb 2.30 Hosp D Friday 16th Feb 1.00am Hosp E *Signed*			

Figure 7.6. Acute unit: team briefing

proportion who feel they are kept informed by their immediate manager increases to relatively high levels (50-80%). This is found, especially among ancillary staff. The systematic format of team briefing results in more regular communication, especially when previously there were no regular management and staff meetings. This method is one of the key ways to ensure the *Heineken concept* operates and reaches the parts (of the unit) that other communication approaches do not reach! (Silverstone 1990).

Open forum meetings

A different form of communication is the large *open forum* meeting in which the chief executive, or senior manager, invites **any** member of staff to attend a meeting to discuss particular issues of topical concern. The meeting usually takes place in a large meeting room, lasts between half to one hour, and has only two rules:

> Staff preferably stay the whole time, rather than walking out
> Participants do not hit each other!

This rather unusual set of rules immediately encourages an open-style of communication, which must itself be modelled by the chief executive in their own style of communication in answer to questions raised. These meetings can benefit from a mini-agenda or topic; or alternatively no topic.

Initial concerns held by potential leaders of such meetings are, for example, 'They will hang me' especially if particularly contentious issues are currently under discussion. As yet this fate has not been heard to have befallen a general manager. The manager with good and robust/assertive communication skills can adeptly channel 'over enthusiastic' communication into useful debate of issues about which staff sincerely hold deep attitudes and feelings.

Unit newsletter and TQM information

Many provider units, and prior to general management, individual hospitals, have successfully operated *Newsletters* which attempt, and often succeed, in achieving several objectives. They effectively:

> Disseminate formal and informal information top-down
> Provide information in an interesting and more entertaining way than more formal documentation
> Provide a method by which staff can communicate with the *whole unit*
> Provide an opportunity for senior management to advertise key unit strategies or plans in an acceptable graphic way.

Communicating interest in quality can also be enhanced by the use of:

> Suggestion schemes
> Induction programmes
> Poster campaigns
> Competitions
> Prizes
> Demonstration and exhibition
> (Oakland, 1989)

Teamwork enhancement

When well established and managed, teams facilitate communication and quality improvement by improving the flow of information and the process of disseminating/sharing good practices as well as improving the process of problem solving. Teamwork is an essential component of TQM implementation (Oakland 1989) in that it builds up trust and interdependence as well as improving communications.

Empowerment of teams is a critical part of the management process. It requires huge effort and commitment by managers to ensure it is put into practice at all appropriate levels. Two particular types of teams aid quality improvement within provider units: quality improvement teams and quality circles.

1. *Quality improvement (QI) teams:* Comprise a group of staff brought together to investigate and solve a particular problem because between them they have a relevant set of skills. As such they will usually be cross-functional and multidisciplinary. Key issues of team selection and leadership, team objective setting, team meeting and assignments are important. However, perhaps the most crucial process is that of team dynamics or the *chemistry* of the team, facilitated by its leader to:
    Create an atmosphere for innovation and creativity
    Encourage full participation by team members
    Allow differences of opinion to be aired
    Encourage *brainstorming,* and occasional irrationality!
    Support the team membership.
  Projects themselves may be suggested for QI teams by senior management; QI teams themselves; *purchasers;* internal customers or suppliers (staff); or external customers: patients or their representatives — GPs.
2. *Quality circles:* Consists of a group of staff who usually work in the same location and meet regularly and voluntarily work to sort out work related problems in their area. They recommend solutions to senior management, often to be carried out, if sanctioned, by themselves. The main advantage of the quality circle is the production of *shop floor motivation* to achieve quality improvement of a highly practical nature. They have been described in depth in Oakland (1989) and Robson (1982).

The team approach to quality improvement is effective: it enhances the potential of staff to problem-solve and harnesses the skills and initiative of those involved. Senior management learn to delegate decision making responsibility and influence to such teams learning that the outcome is well worth the initial *letting go.*

Standards setting, monitoring and review

As has been mentioned at length in an earlier chapter, all quality services and processes can, and arguably should, be *subjected* to the standard setting process. Communication within a unit is no exception. All the foregoing approaches and subsequent behaviour can be operationalised into active standards for effective communication. Once set, unit performance can be measured against these standards and regularly reviewed, updated and improved.

Communication audit: patients

To what extent does a provider unit know and understand how it communicates with its main consumers — the patients. Patients come into contact with several different groups of staff as they *walk through* the service, such as the receptionist, nurse, doctor, radiographer, OT, ward clerk, and so on.

The flow of information and the ease with which this occurs, is a prime component for auditing, within the context of effective communication.

The flow of information is obviously 2-way:

```
1.  Staff  ←—information—→  Patient
                 ↗ feedback
                 involvement
2.  Staff  ——audit——→  Management
```

1. *Information from staff to patients:* Most, if not all, provider unit staff are part down the road to ensuring information which is given to patients is:

    Accurate
    Appropriate to the patient's needs
    Appropriate to their reading ability and ethnic origin
    High quality production.

    The internal audit must clarify the extent to which high quality information, in terms of the five criteria listed above, is consistently provided to patients in *each* patient-contact area.

2. *Customer feedback:* There are three important reasons for finding out what the main customer — the patient — thinks about the service he/she received (Carr-Hill et al 1989):

    a. It identifies customer dissatisfaction and can provide management with information about non-conformance.
    b. It displays a willingness to involve patients in active participation in planning of service delivery/improvements.
    c. It is thought at the time, to boost patient morale and recovery by involving him/her in the treatment process.

Areas worthy of audit are as follows:

1. Communication with patients who complain or make suggestions.
2. Strategy for regular elicitation of customer views on service delivery: surveys; user group.
3. Appropriateness of methods: questionnaire design; use of interviewers; sampling; type of analysis.

Communications improvement implementation: patients

1. Information to patients

Units should promote realistic public and professional perceptions of what they can provide as the best possible affordable service. Information about a local provider which the public obtain from friends, colleagues, and relatives should be complemented wherever possible by information readily made available by the provider unit itself, as promoting accurate expectations among the patients, is a vital part of *customer relations.* Someone who knows how to use a service is more likely to use it, and use it appropriately to greater benefit.

Written information is extensively used throughout the NHS. However, in any one provider unit the extent to which all information given to patients meets the five criteria above is inconsistent and variable. For example, ward A may provide professional (albeit economically) produced admission leaflets to patients, whereas ward B still relies on photocopied and out-of-date information which may be inaccurate.

Information given to patients can also cover several mediums:

Written
Face to face/verbal
Telephone/verbal.

Each of these have their own characteristics needing special consideration in information production and dissemination. The information can also cover any of the main aspects of outpatient, inpatient, and day care treatment that patients may encounter.

Information **provision** is an essential part of customer relations: it is important to provide the required information in a style and format which suits different categories of people.

2. Information from patients

Background literature relating to customer feedback approaches in the NHS has been well documented by Carr-Hill and colleagues (1989) in work carried out in the Centre for Health Economics at the University of York.

A summary of their extensive work indicates that:
- Customer feedback surveys are expensive in time, resources and raising expectations of what can be provided.
- Many surveys are carried out with unclear aims; insufficient resources (including analysis); and little understanding of the complexities of undertaking a survey.

Several other methods, in addition to surveys, can be utilised:

Suggestion cards
Radio phone-ins
Ward meetings: general, therapeutic community, psychiatry
User groups eg outpatients, X-ray, fracture clinic
Informal feedback
Direct patient-therapist feedback.

The key issue, within a TQM framework is: What happens to the information gathered about non-conformance, poor service or potential improvements? Time and energy is increasingly spent by managers in collecting patient information for several reasons:

Managers are *tasked* via IPR to do this
Patients willingly proffer information
Managers have a real interest in learning how to improve their services as a result of patient feedback.

Irrespective of which of these reasons pertain, it is important in terms of how such information, once gathered, is used.

Does the manager get the information analysed and in a form which encourages and facilitates further action to improve the services?
Is this information compared with previously collected information on the same area to highlight trends or insufficient remedial action?

It can be argued that if a manager were only able to utilise **one** process of TQM to improve the quality of service, then customer feedback should be that sole process as it is fundamental to TQM and quality improvement. *Getting close to the customer* is an essential requirement to establishing if the service is meeting or exceeding customer expectations and needs: the primary aim of the service.

Chapter 8 Training for quality — investing for the future

Two well known NHS sayings are:

> Staff are our most valuable resource, and
> Training is the way to change and develop services.

Logically most, if not all, managers would agree with these statements. However, what would vary across all provider units is the commitment and its behavioural and financial implications for training.

Training for quality should appear high on the list of objectives and priorities in the TQM implementation plan, ie it is scheduled at the outset of discussions on quality improvement.

Responsibility for training in quality rests in three locations, although ultimately the responsibility of the chief executive along with every other aspect or function in health care delivery.

> Line managers: to ensure training is planned, resourced and carried out
> Quality assurance officer: to assist in planning the content and application of training
> Training officer: to carry out the training.

Training in quality must satisfy the following requirements:

> Unit level managers must be committed and involved in planning and setting training priorities
> Objectives must be adequately and realistically resourced and timetabled
> Training should be considered for all parts and levels of the provider unit.

In defining the programme, the specific needs of staff for quality training must be identified, for example:

> Which staff need to be trained: group, level, location
> Level of attainment required: awareness, skill, changes
> Length of training: time, duration
> Specified and explicit benefits expected from training
> Type of training and trainer required
> Resources required: finance, manpower, facilities.

The type of programmes and appropriate materials will vary dependent on topic and group being trained. This should **not** be predicted by the skills and predilection of the trainer! Time spent on reviewing critically available programmes or in designing unit-specific materials is time well spent.

Training takes time and the effective implementation of *training for quality* programmes require extensive commitment by managers, trainers, and staff if real behavioural change and organisational development is to occur. As such the effects, real or imagined (!), must be monitored regularly. Benefits should be seen and be measurable, and relatively easily translated or evident in improved health care quality outcomes. These various steps are illustrated in Figure 8.1.

In one sense, a systematic approach to quality training should concentrate on raising awareness and competence/skills in three main areas:

> Prevention of problems and errors

Reporting and analysis of problems and errors
Investigation, modification and learning from problems and errors.

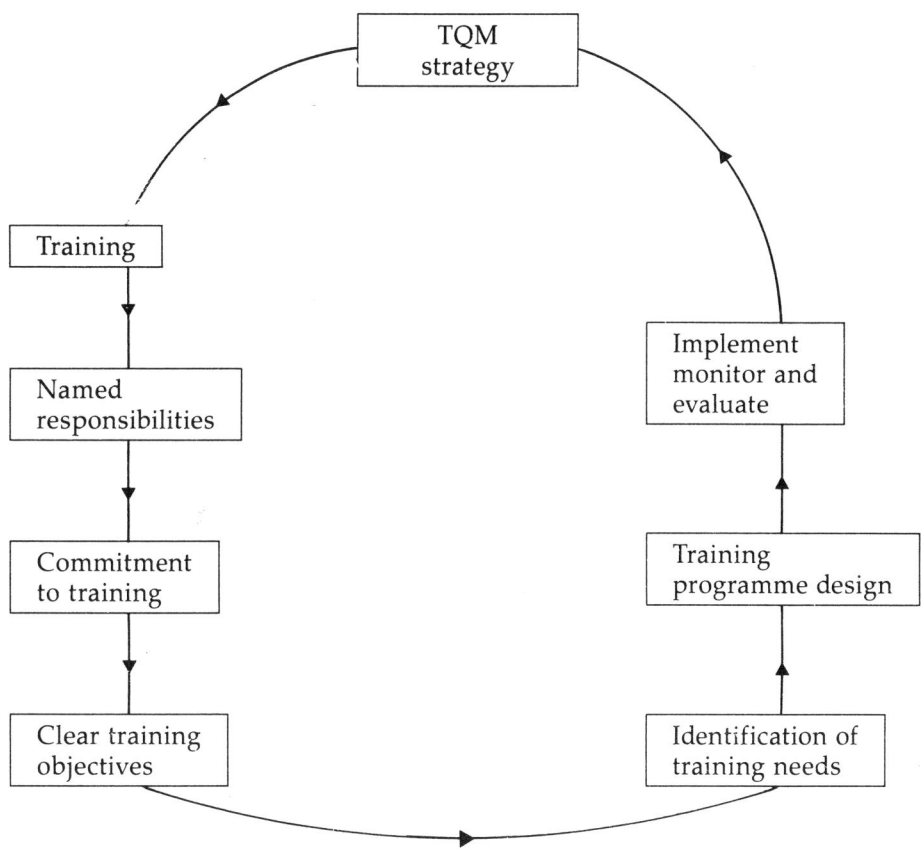

Figure 8.1. Summary of quality training cycle

Prevention of problems and errors

To facilitate training for prevention of problems in a service, the following components are essential:

Information given concerning unit TQM policy; current standards of practice; job description including quality requirements
Understanding and skills in setting, monitoring, and reviewing standards
Understanding and skills in main TQM concepts of consumer sensitivity; service audit; communication; quality costing; resource management.

Reporting and analysis of problems and errors
All staff must be trained or at least made more aware of the importance of recording, and analysing meaningfully the various problems, errors, and apparent inefficiency which may occur in any aspect of health care delivery. As well as reinforcing and practising appropriate corrective action, skills in **learning from** mistakes is essential.

Problem investigation
An extension of the last statement above, is the need for staff to have skills in *audit* to provide important information, with the benefit of hindsight and discussion with peers, which can feedback to enhance prevention strategies.

Training needs

Training in TQM must occur at all levels of the unit as quality improvement is part of everyone's job. Therefore, it should involve:

> Senior management: UGM/CE, Unit board members and clinicians
> Middle management: hospital managers, department managers/heads
> Supervisors and ward managers
> *Front line* carers and support staff
> Trainees on post qualification courses.

Time and commitment spent on providing effective training for the majority, if not all, staff will pay dividends in terms of quality improvements. The reverse will also occur in that the process of TQM will not go naturally or at the pace it should and that the patient deserves if training is not offered.

Can different training needs be identified at each of these levels? Below are some of the main training needs for each of the five levels of staff identified:

1. *Senior management*
   TQM Awareness
   Increasing commitment
   Awareness of potential obstacles and solutions
   Competition
   Patient centredness
   Quality in terms of cost, workload and customer care
   TQM processes:
     Meeting customer requirements: internal and external
     Setting standards for *zero defects* in care
     Reducing quality costs
     Developing quality systems and contracts
     Involving and motivating **all** staff
   Understanding of clinician/general management interface
   Relevance of team building and quality facilitation.

2. *Middle management*
   TQM awareness and understanding of most, if not all, the areas identified above
   Technical skills required to design, implement, review and change quality systems eg standard setting, clinical audit, under their direct operational control
   Ensuring *quality tools* are active and meaningful to them and can be effectively communicated to their staff with ownership developed.

3. *Supervisors' and ward managers*
   Training in principles of general management and TQM
   Understanding of their vital role in the wider unit
   Realisation of existing commitment by senior managers to quality and supporting them
   Understanding of unit's TQM policy and objectives
   Understanding and competence in quality systems
   Identifying and enhancing commitment to TQM at their level.

4. **Front line** *carers and support staff*
   Patients are **more** likely to meet these staff and more frequently. Therefore, their involvement in quality improvement is perhaps most important of all and their training should include:
     Basics of TQM with relatively little *management jargon*
     Clear and immediate relevance of TQM concepts to their everyday work
     Clarification and skills in quality procedures
     Sensitivity to patient expectations.

5. *Trainees on post qualification courses*
   Brief module on TQM
   Foundation for TQM sensitive attitude formation for staff entering the service.

Although the content of training in TQ Management will vary depending on the training needs at different levels of the unit, the basic materials will need to include the following:

1. *Basic information*
    History of TQM (public and private sector, worldwide)
    Description of TQM approach in health care
    Development of TQM organisational strategy
    Development of quality management culture
    Development of specific TQM components:
        Standard setting, monitoring and review
        Patient information and feedback strategy
        Staff information and feedback strategy
        Communications strategy
        Quality costing
        Training for quality
        Clinical and medical audit
        Resource management and quality
        Implications for contracting
    Evaluation of TQ Management initiatives. Facilitation and support requirements to maintain it.
2. *Provider unit quality audit and diagnosis*
    TQM audit implementation
    Staff attitude survey/interviews
    Recognition of excellence and mediocrity
    Audit of management culture
    Assessment of main competitors in health care locally.
3. *Senior management commitment*
    TQM strategy discussions: development of missions statements, core values, and key objectives
    Manager attitude survey/interviewing
    Team building
    External facilitation of change.
4. *Raising staff awareness and unit strategy development*
    Senior staff communications
    Staff awareness of standard setting, communications, customer relations, quality costing
    Development of unit strategy with staff ownership
    Quality issues from key staff
    Integration of TQM with other major initiatives eg Resource Management Initiative (RMI), contracting and service specifications at provider and purchaser levels.
5. *TQM support structure formation and unit action planning*
    Unit quality steering group
    Development of unit action plan
    Structure of quality action teams and departmental specific action programmes
    Standard setting plan
6. *Implementing action plan*
    Clarity of strategy and plan
    Ownership and commitment of all staff
    Maintaining quality 'momentum' for change
    Involvement of clinicians
    Relate TQM to key initiatives of waiting list reduction, workload planning and clinical audit
7. *Training initiatives for TQM*
    Necessity for training and continuing education
    Different methods and techniques for training and quality
    Need for investment in training
    Performance indicators for training
    Training materials

8. *Evaluating and reviewing TQM programmes*
   Evaluation methods and outcome definitions
   Internal and external review
   Evaluation attitude change
   Linking TQM inextricably to health care outcomes
   Ensure *audit — implementation — review* cycle

Training method

Several different *vehicles* for training in TQM currently exist:

Internal: 1. Workshops
2. Seminars
3. Briefings

External: 1. Qualification courses
   Conceptual and information based
   Action-learning and skills based
2. Short courses workshops
   General training methods
   Skills based
   Case study based
   Attitude changing
3. Simulation workshops
4. Training company task force.

Internal training within a provider unit is often carried out firstly via workshops and seminars led by internal or external trainers. Secondly for a variety of reasons some good, some expedient, *briefings* are used to give concise overviews of key TQM issues to middle-line and junior staff. This *quick shot* of TQM information and enthusiasm has greater credibility and learning potential relative to the status of the *briefer* and the higher the quality of the presentation. Short briefing packs have been produced for acute and community/priority services (Price 1990, Koch 1991, SE Staffordshire package *Introducing Quality Management* 1990)

External training opportunities have developed rapidly since the higher priority of QA and TQM activities in the NHS by the NHSME.

1. *Qualification courses* Over the next five years, post qualification courses in quality management in health care will develop. The first such course started last year (1990/91) at the Health Services Management Centre, in Birmingham, leading to either a diploma or masters' qualification, dependent on evaluation. This two-year course is tutored by academic and staff with NHS management experience of implementing TQM projects. It is run on a modular basis thus ensuring that coursework and discussions link in with real life practical experience in the work setting.
   It is likely that the number of these courses will increase and be complemented by management development opportunities using action-learning techniques. The NHS base may well broaden to include other public sector experience and participants.
2. *Short courses/workshops* There are four main health service management centres *servicing* the NHS training and management development needs: Kings Fund Centre, HSMC Birmingham, Nuffield Institute Leeds, and HSU, Manchester. Short courses and workshops in quality assurance are offered with HSMC Birmingham running an integrated modular programme in TQM.

In addition to giving information, sharing experience and enhancing skills, two innovative models of short workshop learning have been recently developed:

1. *Case-study based learning* Consideration has recently been given (Health Care Evaluation Unit, Bristol 1990) to a methodology for using a case-study based learning approach to TQM training. It would involve utilising both real material from one or more TQM demonstration project sites, or fictitious material, which exemplify different TQM processes, operating at different levels of a provider unit, at different stages of implementation and with differing outcomes as illustrated in Figure 8.2.

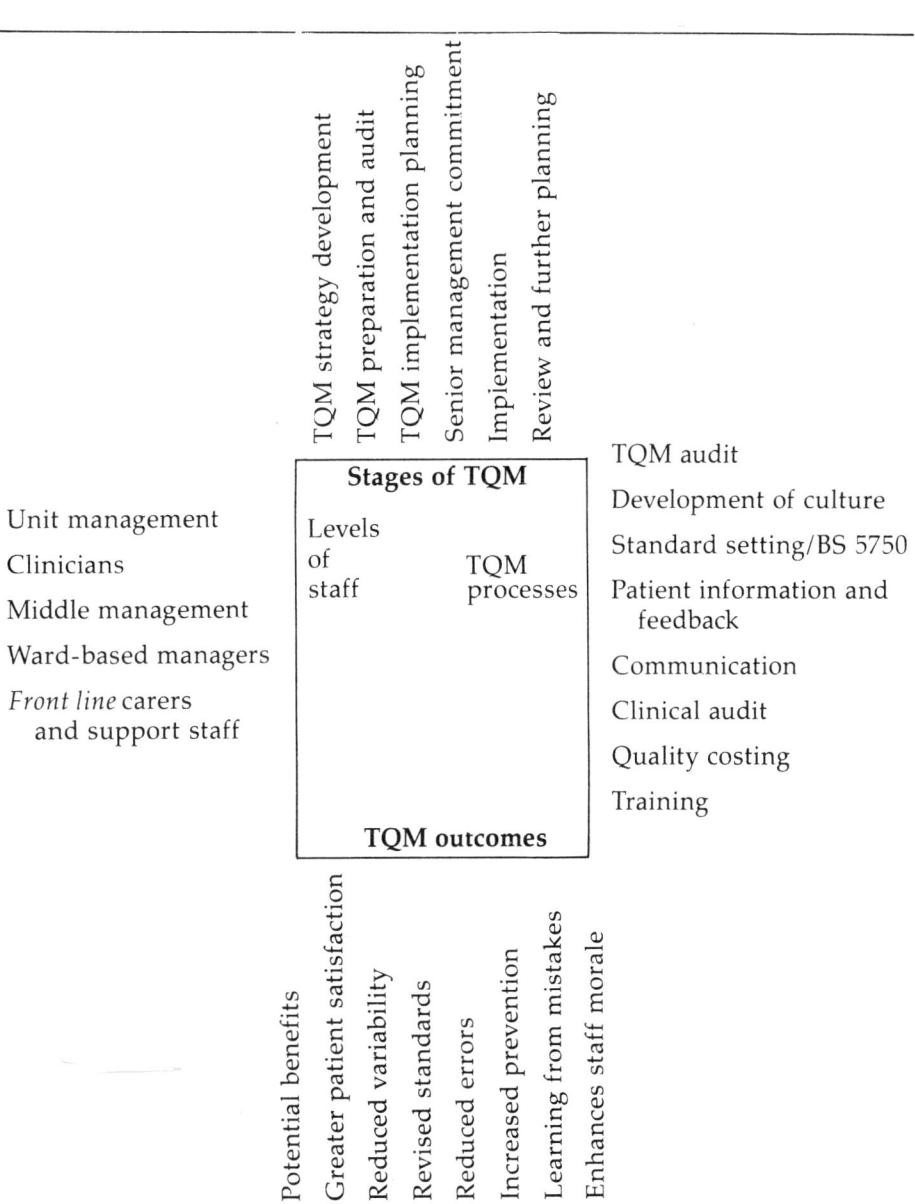

Figure 8.2. Stages of TQM and outcomes

2. *Attitude-changing approaches* Underlying most training programmes is the implicit hope or objective that attitudes of participants may, through being given new and contradictory (to their own) ideas, be modified. A particular approach which can hasten this process is a more explicit way derived from the considerable body of theory and practical research into *personal construct theory* (Kelly 1955, Koch 1981). This approach quite simply elicits from a particular person the key people or *elements* which

are important to them when considering a particular part of their life (eg work, home); or a particular difficulty (eg social, anxiety, assertion); or a particular initiative (eg quality control!). For an NHS manager, looking at his/her attitudes to quality control and improvement, his/her key *elements* might be:

> Self
> Unit general manager/chief executive
> Quality assurance officer
> Consultant on UMB
> Other consultant, particular individual or *caricature*
> District quality adviser
> Ward sister, particular individual or *caricature*
> Patient, particular individual or *caricature*.

This approach then invites the person to compare and contrast groups of two or three of the above people/elements and identify ideas/ concepts/*constructs* which come to mind. In the context of the eight elements above, and the quality issues, typical constructs, and their opposites, might well be:

Constructs	**Opposites**
Influential	Weak
Committed	Pays lip service to quality
Gets things done	Talks about improvement
Sees *quality* linked to *quantity* and *cost*	Sees *quality* as the prime objective irrespective of *cost* and *quantity*
Sees *quality* purely as patient linked	Sees link between patient-care and organisational culture
Practical	Theoretical
Team orientated	Independent

Through a simple process of rating *elements* on *constructs* (eg self scores 8, on a 10-point rating scale, on 'influential — weak' construct), a composite picture or *grid* can be constructed which illustrates how each participant sees himself/herself in relation to key colleagues; and also identifies the two main dimensions of attitudes which appear pertinent to the way he/she sees his/her *world* or at least the world of quality! Figure 8.3 is an example of such a grid which recently emerged from a workshop, for quality assurance managers held by the author in Birmingham (1990). In fact, it is a composite group grid which indicated a high level of concordance between the group on this topic.

Following the development of such a grid, the ensuing discussion can itself result in small attitude changes. For example, one key issue in this particular group centred on whether the QA manager had, in fact, more **real** power than was felt through the advisory role without line management responsibilities and restrictions. This potential change in attitude can be further enhanced through skills training and role playing by simulated situations where quality improvements can be influenced successfully. Fixed role playing and grid construction are eminently applicable and proven approaches to attitude change in quality training.

3. *Simulation-based learning* TQM simulation takes staff through the key elements of the new provider unit-driven quality conscious era and, using either method described below, covers the key components of implementing TQM. The two methods include:

either — using case study material,
or — real life *back home* material

to establish a simulated version of a typical provider unit attempting to implement TQM in the context of:

Recurrent financial pressures
Other major initiatives eg RMI
Integration with emerging contracting process
Moderate ambivalence to *business* concepts of TQM and *customers*

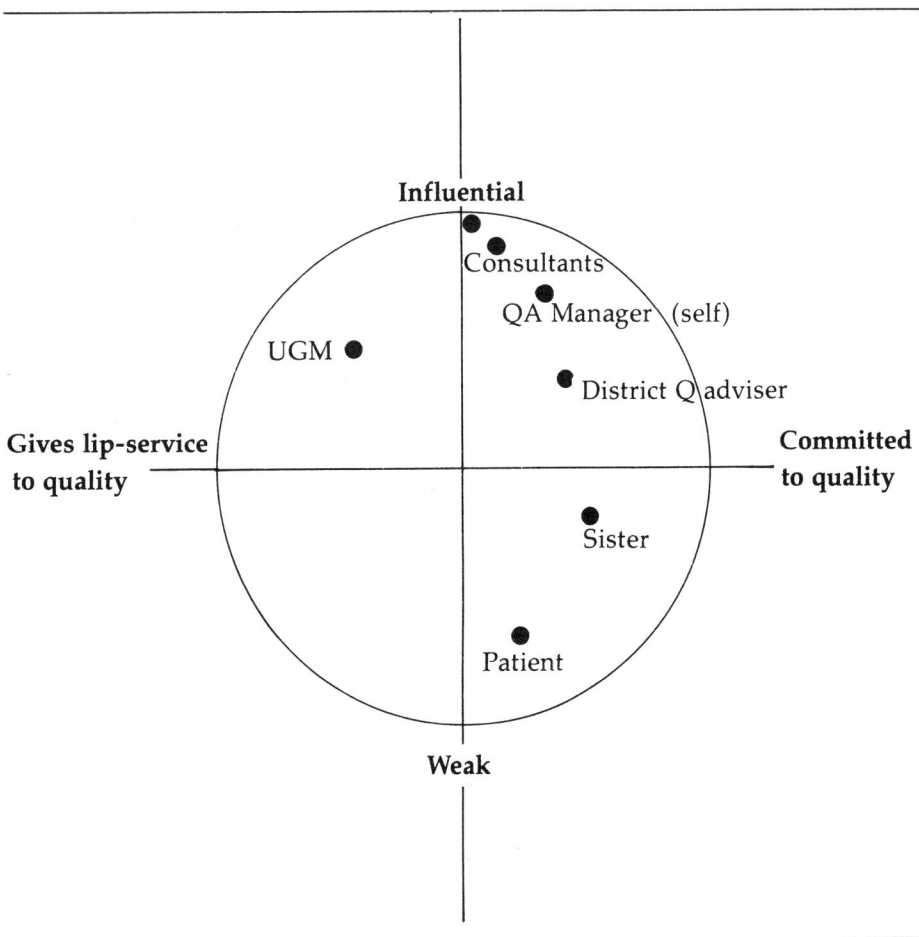

Figure 8.3. Grid to show relationship with key colleagues

Note: Training needs to address each side of the grid in its focus. Raising awareness of: levels, outcomes, processes, stages.

The package includes a specially designed computer software model enabling participants to test the impact of different options and take away for use in developing their own business plans.

The aim of this approach (Koch and Trembley 1991) is to improve understanding of how different interventions under the TQM *umbrella* can affect the operations with a hospital or community service. Simulation allows full consideration of the effects of interventions on patient services in as near a real life/operational situation as possible.

4. **Training company** *task force* An innovative approach being developed by the author centres on a short-term project-based task force set up for a period of one to three years linking two to four provider units with a training consultant, who may be affiliated to a higher education establishment and is modelled, in part, on the Teaching Company Scheme (TCS) developed in Swindon (1987) for use in industry.

Its aim is to foster and harness the management and academic skills to produce short-term improvements in health care performance. The project team is an instrument of change and provides an active and continuous partnership between the training consultant and the practitioners in the provider units with the prime objective of raising the level of health care performance by effective use of staff and other resources. In addition, such a strategy achieves the following objectives:

Improvement in management approach
Implementation of advanced quality schemes
Develop and retrain existing hospital and community staff
Enhance perceived relevance and context of *in-house* training and consultancy.

The agents of change are the training consultant *and* the two to four provider unit *trainees* who themselves are qualified professionals wishing to extend their experience of TQM **simultaneously** with the growth of TQM in their unit. The process involves:

Short workshops in each of the provider unit locations for all four participants, led by the training consultant covering key aspects of TQM implementation
Series of consecutive days working on TQM project implementation **as a team** in each of the provider units, utilising the knowledge, experience and group-enhanced creativity of *all* members of the team, led and facilitated by the trainer.

The outcomes of this programme are several:

High quality practical TQM experience for each participant in terms of information, skills in TQM system, implementation management experience
Provider unit TQM strategy development and implementation facilitated by a well briefed and enthusiastic project team whose practical experience exceeds not only that of the trainer but also the *sum of the team's parts*...
Excellent management and career development for the participants.

The success of this approach is critically dependent, as it was in the original TCS Scheme (1987), on people, relationships, and commitment not only of the individuals, but also of the *host* organisations/provider units. The participants must be *recruited* for the correct blend of personality, credibility and experience.

Perhaps TCS can turn TQM in the NHS into TQC — Total Quality Care!

Conclusion

Managers and staff in the NHS must be equipped with new skills, such as TQM, to meet the contemporary demands of the public for higher quality, increasingly patient centred health care. Training should have coherence and wherever possible be multidisciplinary, reflecting the usual way that patients experience care. It must also be integrally linked to the unit's business plan and quality strategy.

Most staff enjoy training, when offered the chance. However, most staff and managers who fund training are ambivalent about whether training translates into organisational or service change. Training outcomes can be of four kinds of change:

1. Staff awareness and competence/skill
2. Staff **confidence** and morale
3. Quality of **service** provided to patients or internal customers — other departments
4. Health of patients eg surgical audit training and readmission rates

Most training events would achieve outcome 1 — Awareness raising. Clear objective setting for training and management commitment to support the *trainee* via explicit agreement as to why training was necessary not only for the *trainee* but also for the unit, and also by showing interest and support to implementing newly acquired skills once the *trainee* returns to the unit.

The final point to be made has to relate to *costs of poor quality* mentioned in chapter 2 as part of auditing poor performance. In training terms, this cost rests with the poor quality, relatively speaking, that is perpetuated by **not** investing heavily enough in training the *most valuable resource.* Compared with other large companies in the private sector, with equivalent budgets; span of operational control; and perceived importance to the community; the NHS does not invest enough as a proportion of its revenue budget in training in general and in *training in quality* in particular. Focused, evaluated training is worth having and funding!

Chapter 9 Evaluating the contract for quality

The process of organising and managing quality improvement in health care delivery — namely, TQM — must be seen as an integration of the several recent key strategies which have been adopted in various units in different ways. Strategies such as general management, resource management, clinical directories, quality assurance, medical and clinical audit, and business planning are all complementary to TQM in their implicit or explicit objective to offer increasingly high quality care to the patient. The recent legislation has now added the additional strategy of *contracting* which supposedly will provide a *competitive* edge to the organisation of health care and allow the patient and their GP more realistic choice. This awaits the test of time! Most general managers are developing their knowledge and skills about integration and complementary strategies, followed by consideration as to how these many plans fit together. TQM offers, perhaps, the most realistic, practical and successful model by which these plans can be integrated within a contracting environment.

TQM has to be seen as a process, not an *off-the-shelf* programme of short life (Crosby, 1984). If the benefits which can be achieved through the implementation of a *total quality* approach are to be realised, the approach must gradually become permanent within the culture of the service. TQM must be seen as the *right way to manage* with continuous improvement being the order of the day in **all** areas. This ongoing *maintenance* of the total quality approach has been cited in many private companies as the predictor of success. It is the easier of two large tasks to **establish** a TQM plan — to sustain it for several years is much more difficult. For this to occur, managers and clinicians alike will require vision and imagination to marry up short-term gains with medium and long-term objectives, for gradual improvement in health care delivery. This vision must include the realisation of how important it is to reduce *process variability* ie to continually search for and reduce, sensibly, activity in which staff or technical processes are variable for no valid or predicted reason. This will lead to increases both in quality and *productivity* of health care.

As a focus for strategies of health care delivery, TQM must ensure that the key contractual obligations to provide services are met:

> Accessible
> Effective
> Acceptable
> Appropriate

Services are organised with appropriate input of:

1. Clear management commitment from both clinicians and general managers; leadership; and capabilities
2. Optimum team work and recognition of staff value
3. Implementation of quality techniques
4. Monitoring and identification of performance against contract specification and reduction of *non-conformance.*

Throughout the service:

> Unit/trust
> Directorate
> Speciality
> Individual members of staff

These areas are complementary and illustrated in Figure 9.1.

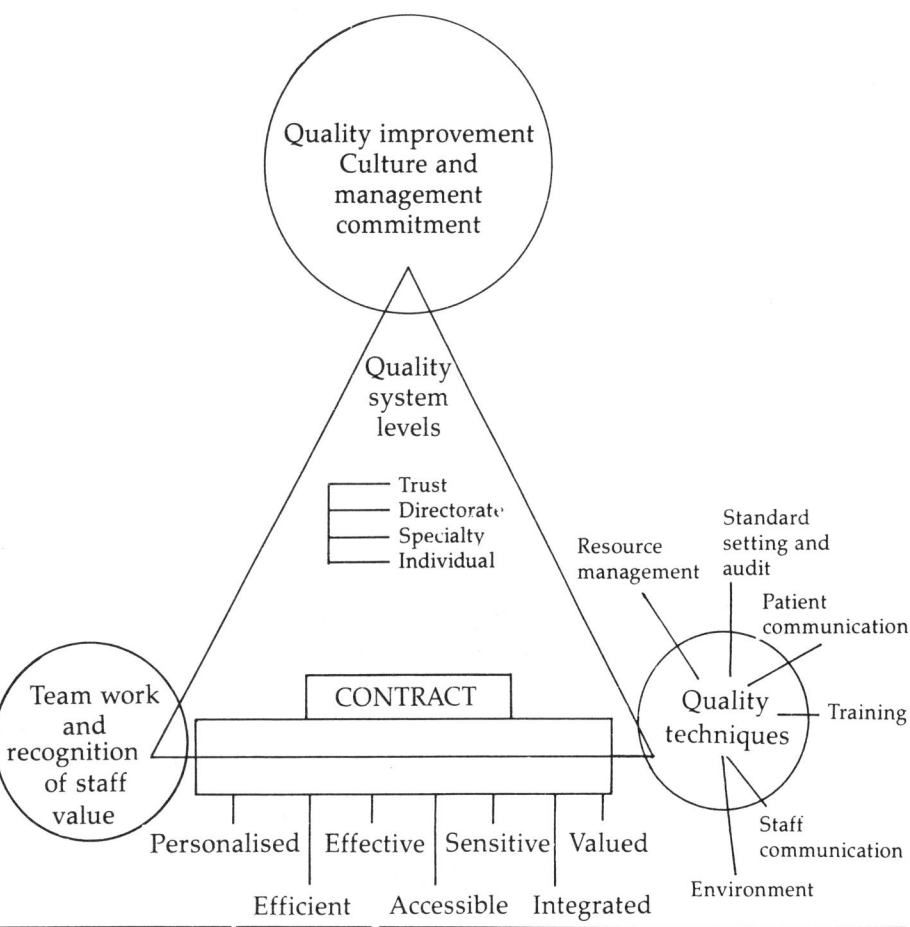

Figure 9.1.

Within the provider unit, as contracts focus directly on specific clinical services offered, the key mechanism for achieving the integrated *meeting* of contract specification is via the specialty level *Service/Business/Quality Plan* which facilitates operational monitoring and marketing analysis by clinicians and general managers together. It is important that clinicians rapidly identify and agree, corporately, specialty by specialty, on basic indicators of quality of care and suggest to senior management that their view as medical staff should be the foundation of the service specification, in quality terms. In specific terms, the key benefit of effective health care delivery is that in meeting contractual obligations patients receive equitable, relevant, and high quality service. In general, there are internal service benefits of TQM that contributes to this overall success. For hospitals and community services which have started on the *total quality* road, the results will be steady improvements and reductions in waste and unnecessary cross-boundary flow of patients. The benefits are many:

- Improved service image — From a mediocre, *they do their best* service to a publicly reinforced service that puts the customer first
- Improved throughput of patients — Productivity is about eliminating *bottle necks* and disruptions in the patient trail through the service
- Cost reductions — Quality costing followed by quality improvement strategies **saves** money as well as improving care

- Reduced errors and inefficiency — By bringing **all** care and service processes under scrutiny and control will reduce errors and inefficiency and make processes quicker and more flexible to changing customer requirements
- Increased consistency of excellence and good practice — Variable practice of variable quality can change towards greater consistency amongst wards, teams or departments in terms of clinical and non-clinical practices
- Improved management — TQM promotes better management, based on clearer goals and objectives, improved decision making and higher prioritising of quality issues
- Satisfied patients — The more these other benefits are realised, the more customers are and feel they are put first, the more they will feel satisfied.

In many cities, patients and GPs will increasingly have the opportunity to *look around* at different services for the help they require. They will, undoubtedly, obtain impressions of what units can offer in terms of initial *helpfulness*. Choice will be made partly on this basis. Services, therefore, will need to be more committed to improving their *consumer sensitivity*, finding out what patients and their GPs want and expect. During the delivery of care, attention must be given to:

1. Helping patients make the best of the occasional inevitable bad experience in hospitals.
2. Showing concern for others (indirectly) rather than indifference.
3. Encouraging patients to have the *most* positive experience with optimal outcome.
4. Unsolicited *giving* by staff.

The overall constructive self-criticism of effective personal contact, in day-to-day patient care, forms a key attitude of most NHS staff and is fundamental to the continuation of TQM culture. How will TQM in health care develop over the next five years? It is likely that the initial pump-priming, development funds from the Department of Health, started in 1989, will cease, putting the onus on purchasers and providers to negotiate appropriate pricing of services and developments to allow providers to finance TQM initiatives, over and above sustaining low-cost quality improvement activities. The TQM techniques and systems which have been discussed in previous chapters will require continued development and refinement to ensure that cultural change translates into meeting contractual obligations. Particular areas which will bear fruit are:

1. *Statistical process control:* As a method fundamental to successful TQM programmes in industry this method of identifying and controlling the variability of service provision which has not been applied to health care merits testing.
2. *Quality costing:* In general, a greater awareness of staff, responsible for providing care and service to patients, will facilitate a reduction of the costs of poor quality and will, therefore, contribute, in part, to desired developments.
3. *Purchaser/provider strategy for negotiating quality improvements via contracting:* In the early part of 1991/92, both purchasers and providers will be *mimicking* the market approach to contracting for services. During

this time, the positive *win-win* skills which enable both providers and purchasers to obtain desired improvements in delivery of care and organisational benefits, will develop.

The National Health Service still remains the best organised service for delivering health care in the world and the most efficient and effective of this country's public sector services. Having said this, *total quality* has in no way been achieved in any provider unit, as yet. However, moving from this position of strength and confidence, staff and their management/clinician leaders will find that total quality processes will improve services considerably and help patients receive increasingly higher quality of care, for which they will feel even more satisfied.

Chapter 10 References

Bagust A 1989 Resource management or managing resources. *Health Services Management Research* **2, 3**: 217-220

Bennett J. Walshe K 1990 Occurrence screening as a method of audit. *British Medical Journal* **300**: 1248-1250

Black N 1989 Quality of care in the UK. *International Journal of Health Care Quality Assurance* **2, 2**: 20-22

Carr-Hill R. McIver S. Dixon P 1989 *The NHS and its customers.* York Centre for Health Economics. University of York

Crosby P B 1979 *Quality is free.* McGraw-Hill. New York

Crosby P B 1984 *Quality without tears.* McGraw-Hill. New York

Demming W E 1982 *Quality, productivity & competitive positions,* MIT. Cambridge, Mass

Demming W E 1986 *Out of crisis,* MIT. Cambridge, Mass

Diskern S. Dixon M. Halpern S. Shocket G. 1990 *Models of clinical management.* IHSM. London

DTI 1989 *The case for lasting quality.* Caldwell Price. London

DHSS 1969 *Report of the committee of inquiry into allegations of ill-treatment of patients and other irregularities at Ely hospital, Cardiff.* **CMND 3975.** HMSO. London

DHSS 1972 *Management arrangements for the reorganised NHS.* HMSO London

DoH 1990 *Working for patients.* HMSO. London

Feizenbaum A V 1988 Total quality developments into the 1990s in Chase R L (ed) **TQM.** IFS Publications. New York

Freeman A 1990 Information — Lifeblood of the NHS. *Health Services Journal,* **1, March**: 323-334

Health Care Independent 1990 Action on Audit. Proceedings of Conference. London

Hendrick T E 1988 Pre JIT/TQC. Audit: First Step of the Journey. In Chase R L (ed) **TQM.** IFS Publications. New York

HMSO 1990 Framework for information systems: working paper II. *Working for patients.* London

Hopkins A 1990 *Measuring the quality of care.* Royal College of Physicians. London

Ivey A 1983 *Intentional interviewing and counselling.* Brooks Cole. California

Jones S 1990 Medical Audit (Surgery). In *International Journal of Health Care and Quality Assurance* **000**: 00-00

Juran J M 1979 *Quality control handbook* 3rd edn. McGraw-Hill. New York

Juran J M. Gryna F M 1980 *Quality planning and analysis* 2nd edn. McGraw-Hill. New York

Kelly G A 1955 *The psychology of personal constructs.* Norton. New York

Keyser 1989 ODI Survey *Total Quality Management.* **Feb:** 110-115

Koch H C H 1988 *General management in the health service.* Croom Helm. Beckenham, Kent

Koch H C H 1990 Buying and Selling High Quality Health Care in P Spurgeon (Ed). *The changing face of the NHS in the 1990s.* Longman. London

Koch H C H 1990 *TQM training brochure.* Cheltenham and District Health Authority

Koch H C H 1990 *Methodology for Case Study-based Learning Approach to TQM Training.* Health Care Evaluation Unit, Bristol

Koch H C H. Trembley M 1991 *Simulated Learning Approach to TQM Training* (in preparation)

Koch H C H 1991 *Training Manuals in TQM.* Pavilion Publishing, Brighton

Lewis M 1990 Nursing Audit *International Journal of Health Care and Quality Assurance* February (in press)

Mason A 1990 *Enabling Clinical Work in the South West*. South West Regional Health Authority, Bristol

Mathew D 1990 *Management team effectiveness*. GMTS Scheme. London

Maxwell R 1984 Quality assessment in health. *British Medical Journal*. **13**: 31-34

NHS Management Executive 1989 *Sunday Times* Winning Entries for Hospital of the Year. IHSM. London

NHSME 1990 *Nursing Care Audit*. Department of Health, HMSO. London

NHSTA 1989 *Extension of Resource Management*. Bristol

Oakland J 1988 *Total Quality Management*. Department of Trade & Industry. London

Oakland J 1989 *TQM*. Heinemann. Oxford

Øvretveit J 1990 *Quality Health Services*, BIOSS. Brunel, Uxbridge

Peters T J. Waterman R H 1982 *In search of excellence — lessons from America's best-run companies*. Harper & Row. New York

Price M 1990 *Staff Briefing Pack*. Cheltenham & District Health Authority

RCP 1989 *Medical Audit: A First Report*. London

Robson M 1982 *Quality Circles*. Bower. London

Rooney E M. Wilson R S E 1990 A Quality Management System for an Outpatient Clinic and Service Department *Managing Service Quality* (In preparation)

Sale M 1989 Participation in standard setting. *International Journal of Health Care and Quality Assurance*. **2,1**: 31-33

Seddon Y. Jackson S 1990 TQM and Culture Change. *TQM Magazine*, **Aug,** IFS Bedford

S E Staffordshire Health Authority 1980 *Introducing quality management*. S E Staffs Health Authority

Silverstone R 1990 Team spirit on trial. *Health Services Journal*, **Apr 19**: 590-591

Simpson J 1990 *Resource management and quality*. NHSME RMI Unit Report. Department of Health, London

SWRHA 1989 *Regional approach to medical audit*. Bristol

SWRHA 1990 *Enabling clinical work in the south west*. **Mar.** South Western Regional Health Authority

Shaw C D 1986 *Introducing quality assurance*. King's Fund Paper. No. 64

Stewart R 1987 *DGM's and quality improvement*. Templeton Series NHSTA. Bristol

Turvill T 1989 *Resource management — changing the culture*. NHSTA. Bristol

Wilson C R M 1987 *Hospital wide quality assurance*. Saunders. Ontario